Download the TotalAR app to acce including updated NCEES *PE Civil Reference Handbook* page numbers for the solution references.

Scan this QR code with a smartphone or tablet to download the TotalAR app.

Unique Access Key:
GUPFUCAT

Certain features available through TotalAR targets are provided either on a complimentary basis, for a limited period of time, or for purchase, and in all cases are for the sole use of the original purchaser of this book. The unique code shall not be sold or otherwise transferred to any other person or entity. EduMind, Inc. and its School of PE division have the right to terminate access to any or all of these features for any reason and at any time with or without notice. EduMind, Inc. will make its best effort to resolve any technical issues as quickly as possible, but there will be no refunds, extended access periods, or other compensation for any technical issues, unavailability, or if the complimentary access is terminated for any reason. All content available through the TotalAR app is owned or licensed by EduMind, Inc. and subject to all applicable copyright, trademark, and other intellectual property laws and may not be copied, reproduced, or otherwise disseminated without written permission from EduMind, Inc. To request permission, email permissions@edumind.com.

School of PE
A Division of EduMind

www.schoolofpe.com

PE Civil: Transportation Practice Exam & Solutions

EduMind, Inc. is not affiliated with the National Council of Examiners for Engineering and Surveying (NCEES) and all questions are not actual exam questions or questions provided by NCEES, but are similar in nature to the type of questions expected on the exam. All questions and provided solutions are not guaranteed to be error free and the sole intended use of the questions and solutions is for use as a study aid and not for any practical application.

For more details on the NCEES PE Civil exams, visit https://ncees.org/engineering/pe/civil-cbt/.

Copyright © 2022 EduMind, Inc.

All rights reserved.

EduMind, Inc. owns the copyright on all content, unless otherwise noted. No part of this book may be scanned, uploaded, reproduced, distributed, or transmitted in any form or by any means whatsoever, nor used for any purpose other than for the personal use of the original purchaser of this copy, without the express written consent of EduMind, Inc. To obtain permission, contact permissions@edumind.com.

April 2022

ISBN 978-1-970105-48-3

Printed in the USA

School of PE
An imprint of EduMind, Inc.
425 Metro Place N, Suite 450, Dublin, OH 43017
www.schoolofpe.com

Introduction

School of PE takes test preparation seriously, and we want users to pass their PE Civil exams. This *PE Civil: Transportation Practice Exam & Solutions* volume contains features to help users gain an edge in preparing for the PE Civil: Transportation exam.

A big part of preparing for the PE Civil exams is learning to navigate the NCEES *PE Civil Reference Handbook (PECRH)* and the approved codes and standards for each exam—the only resources test-takers can access during the exam. In most of this volume's solutions, you will see an NCEES *PECRH* section reference, a code reference, or a standard reference, directing the user to the relevant resources and sections that cover the question topic. (The questions that don't have a reference were deemed by our subject-matter experts to be basic knowledge for PE Civil test-takers.) You will notice that we did not include *PECRH* page numbers in the volume, as these change with each new version of the handbook. By looking for the necessary information in the *PECRH* and the approved codes and standards, you will learn how they are organized, saving you time during the exam. To find the page numbers of most of the references that appear in this book, use the augmented reality (AR) feature—*PECRH* Content Locator. The page numbers will be updated with the latest version of the handbook so you will always have the most up-to-date information handy.

To access this feature, download the TotalAR app (see the first page of this volume). This will allow you to gain access to the book's enhanced content. When you see a TotalAR (TAR) code, like the one for the *PECRH* Content Locator (p. iv), simply scan it with the TotalAR app to access the desired content. When you scan your first TAR code in the book, you will be prompted to enter a unique access key, which you will also find on p. i.

With this book, you also have access to an online version of this practice exam via the School of PE Exam Simulator, available on www.schoolofpe.com. If you purchased the book through our website, you should have received an email with a user ID and password. Once you log in, click the icon for this book to gain access to the online version of this practice exam. If you purchased the book from another site—for example, Amazon—please email your name and unique access key (from p. i) to info@schoolofpe.com to gain access to this exclusive online feature.

A Final Note

We have tried our best to make this book error free; it has been technically reviewed, edited, and tested. However, we are only human, and errors can happen. If you spot one, please notify us so we can improve the content. To report an error, scan the Feedback code (below) with the TotalAR app, or visit this feedback page:
https://publications.schoolofpe.com/books/pe-civil-transportation-practice-exam/feedback.

We will verify the correct information and add it to the errata page, which will be made available immediately via the TAR code below and the following link:
https://publications.schoolofpe.com/docs/pectpes/v1/errata.pdf.

We welcome your feedback and want to hear how well this book prepared you for the PE Civil: Transportation exam. If you have any suggestions or comments, please send them to us at publications@edumind.com.

Best of luck!
The School of PE team

***PECERH* Content Locator**   **Feedback**   **Errata**

PRACTICE EXAM

PRACTICE EXAM

1. The cross section shown below belongs to a 650-ft long concrete channel. The concrete thickness is 8 in. What is most nearly the volume of concrete needed to construct the channel given a 7.5% waste?

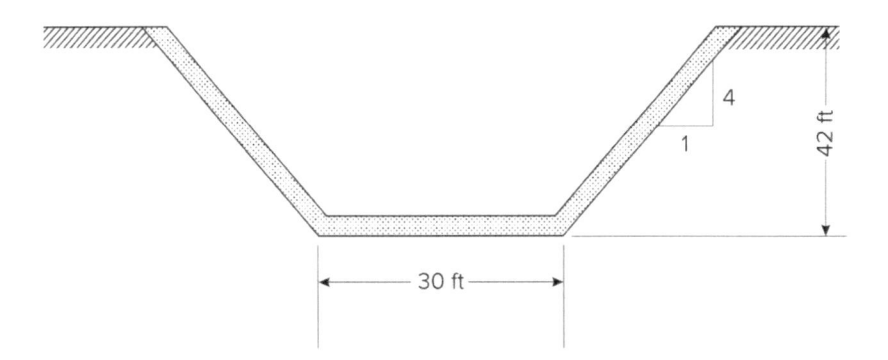

A. 1,265 yd³
B. 1,871 yd³
C. 2,012 yd³
D. 2,219 yd³

2. In the activity network below, what is the total float of task 5?

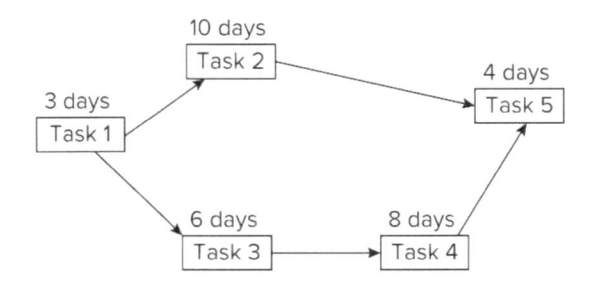

A. 0
B. 1
C. 2
D. 3

3. The owner of a proposed school would like to develop a 100,000-ft² building with enough parking to service staff members and parents. The budget for the project is $2,000,000. Based on previous projects, the owner has forecasted that each parking spot will cost $4,500, and every 90 ft² of building will cost $360 to construct. Assuming the entire building will be constructed, she will be able to afford _____ parking spots.

Fill in the blank.

4. An activity-on-arrow network is shown below. The variables are defined as follows:

TF = total float
LF = late finish
LS = late start
ES = early start
EF = early finish
D = duration

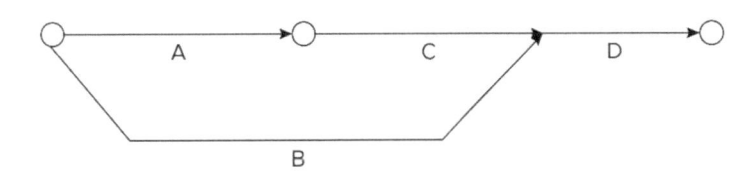

What is the expression for the total float of activity C (TF$_C$)?

    A. TF$_C$ = LF$_D$ − EF$_C$
    B. TF$_C$ = LF$_C$ − ES$_C$ − D$_C$
    C. TF$_C$ = LS$_C$ − EF$_A$
    D. TF$_C$ = ES$_D$ − ES$_C$ − D$_C$

5. A team has 15 days working 8-hr shifts to haul 182,212 yd³ of excavated loose material using a fleet of specified dump trucks that has a capacity of 20 yd³. The total cycle (load, haul, dump, and return) time is 40 min. The minimum number of trucks required is most nearly _____ trucks.

Fill in the blank.

6. A laborer is only able to apply a maximum force (P) of 50 lb applied perpendicular to the top of the wrench. The bolt will only be loosened at a torque of 750 in·lb along the vertical axis in the configuration as shown below. The minimum offset length (L) of the wrench that the laborer must select to loosen the bolt is most nearly:

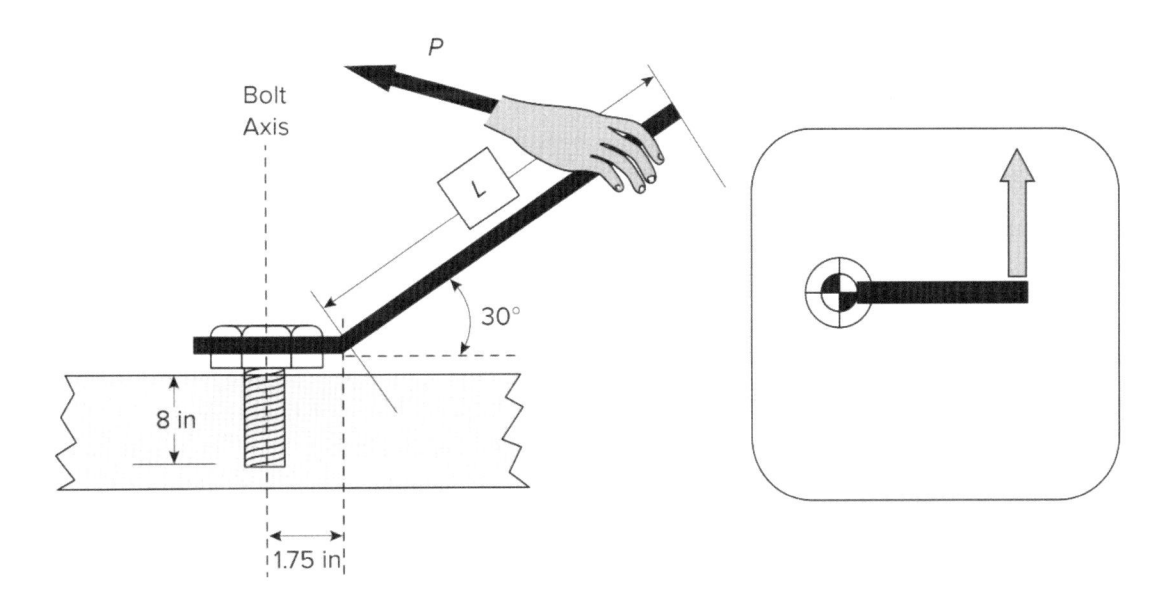

A. 13.25 in
B. 14.78 in
C. 15.3 in
D. 16.15 in

PRACTICE EXAM

7. A formwork system for a cast-in-place wall is braced to support against possible wind loads as shown in the figure below. The connection at the base of the form panel can be considered a hinge. If each brace spaced 7 ft apart can support a maximum compressive axial force of 4,158 lb, the maximum wind load that can be supported is most nearly:

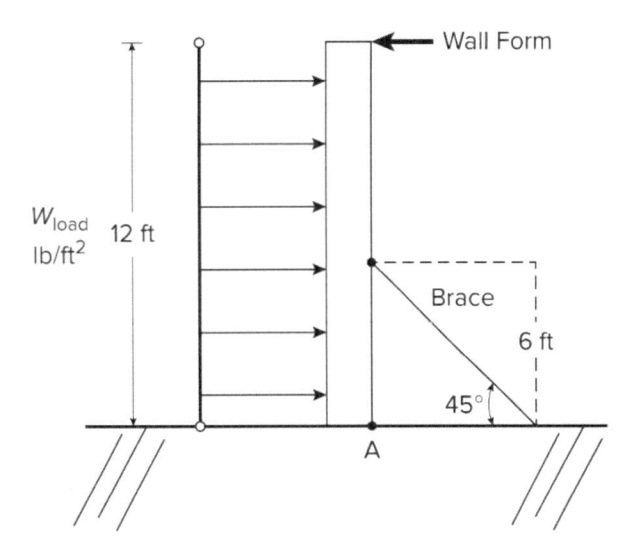

A. 18 lb/ft²
B. 30 lb/ft²
C. 35 lb/ft²
D. 40 lb/ft²

8. The following diagram shows a uniform soil deposit with information about each layer. Assume the effective stress at point A is 90 kPa. Determine the value of H_1.

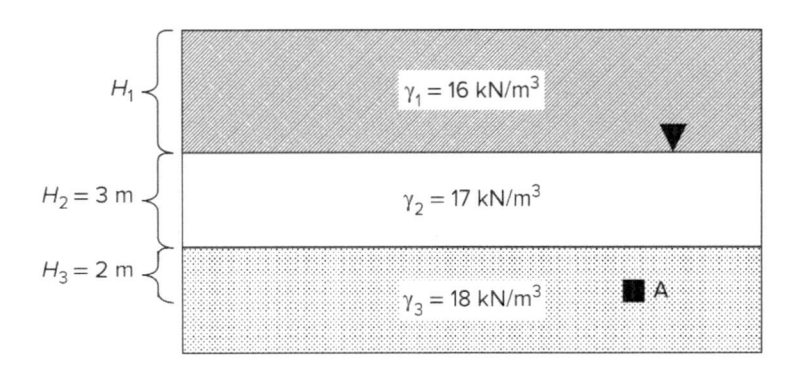

A. 2.3 m
B. 2.5 m
C. 3.3 m
D. 3.5 m

PRACTICE EXAM

9. Point A sits in a sandy soil 8 m below the ground. The underground water level is 3 m below the ground. If the specific gravity (G_S) of soil solids is 2.7, the degree of saturation (S) is 0.5, and the water content of soil (ω) is 20% above the water level and 35% below the water level, determine the effective vertical soil pressure at point A.

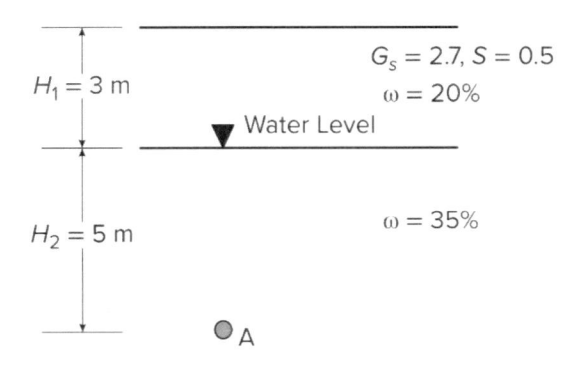

A. 23.5 kPa
B. 35.2 kPa
C. 53.2 kPa
D. 88.7 kPa

10. The following figure shows a 15-ft high frictionless wall. Assume failure along a plane oriented 60° from horizontal and a soil unit weight of 115 lb/ft³. Determine the active earth pressure per unit length of wall according to Rankine theory.

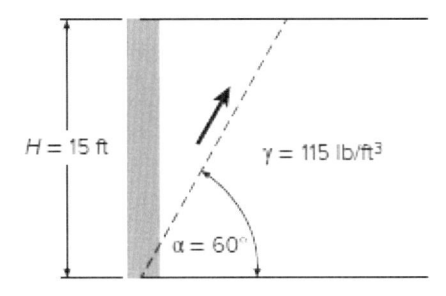

A. 850 lb
B. 4,270 lb
C. 8,540 lb
D. 13,830 lb

PRACTICE EXAM

11. A 20-ft soil sample is assumed to be in the range of virgin compression. Assume the initial effective pressure (p_0) is equal to 3,000 lb/ft², the effective stress increase (Δp) is equal to 2,000 lb/ft², and the sample experiences 12 in of settlement. If the initial porosity (n) is 0.5, what is the coefficient of consolidation (C_c)?

  A. 0.45
  B. 0.68
  C. 0.73
  D. 0.84

12. A 5-m soil sample is given. Assume the coefficient of reconsolidation (C_R) is given as 0.1 and the coefficient of consolidation (C_c) is 0.6. The initial effective pressure (p_0) is equal to 300 kN/m², the effective stress increase (Δp) is equal to 200 kN/m², and the preconsolidation pressure (p_c) is 400 kN/m². In order to determine the initial void ratio, 100 cm³ of initial soil is selected and indicates that there are 45 cm³ of voids. What is the settlement of the soil?

  A. 20 cm
  B. 25 cm
  C. 30 cm
  D. 35 cm

13. The following diagram shows a uniform soil deposit with information about each layer. Assume that the surcharge acting at the top of the soil is 630 lb/ft². If the effective pressure at point A is 3,800 lb/ft², what is the effective density of the second layer?

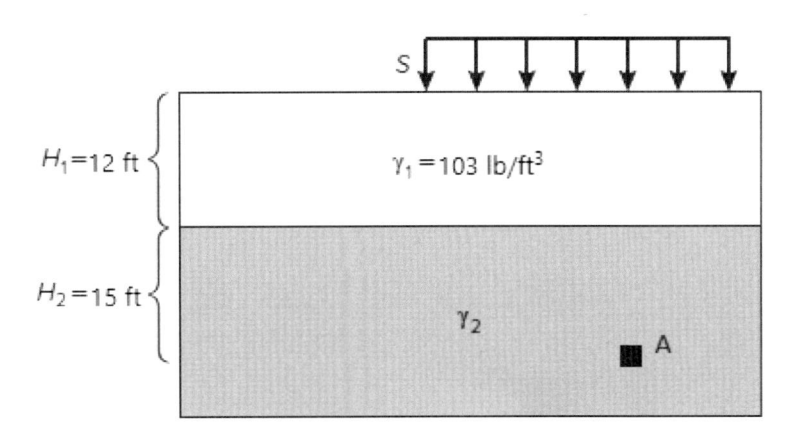

  A. 96 lb/ft³
  B. 117 lb/ft³
  C. 122 lb/ft³
  D. 129 lb/ft³

PRACTICE EXAM

14. Find distance x from the right support B at which the shear force is zero. The beam is 20 ft long as shown. Assume that $w = 10$ kips/ft and $b = 8$ ft.

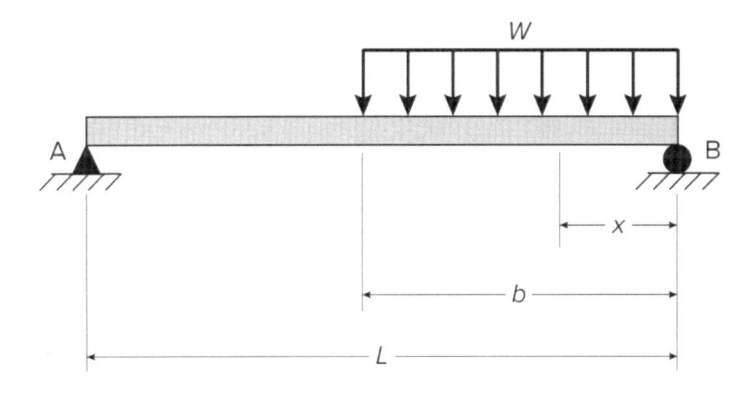

- A. 3.2 ft
- B. 5.6 ft
- C. 6.4 ft
- D. 7.9 ft

15. If an inclined partition wall on a beam has a distributed load of $w = (10 + 10x)$ kN/m, where x is the distance from the left support, determine the amount of shear (V) and moment (M) at point O shown below. Assume that $R = 200$ kN and $L = 10$ m.

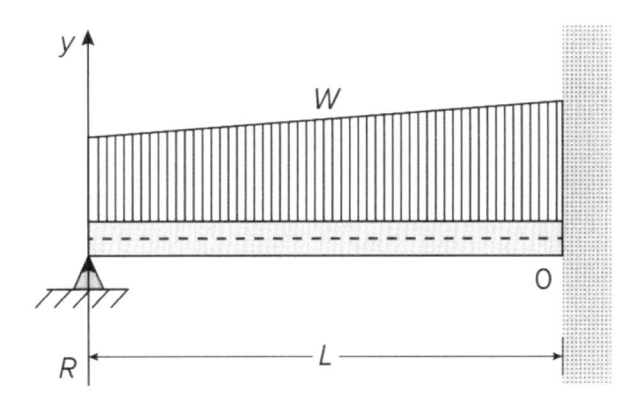

- A. $V = 100$ kN, $M = 26{,}666$ kN·m
- B. $V = 250$ kN, $M = 4{,}333$ kN·m
- C. $V = 400$ kN, $M = 166.7$ kN·m
- D. $V = 500$ kN, $M = 3{,}333$ kN·m

PRACTICE EXAM

16. Assume that $F = 800$ kips, $\theta = 30°$, $a = 6$ ft, $b = 3$ ft, the cross-sectional areas of parts A and B are 5 in² and 10 in² respectively, and $E = 29,000$ ksi. What is the elongation in the system below?

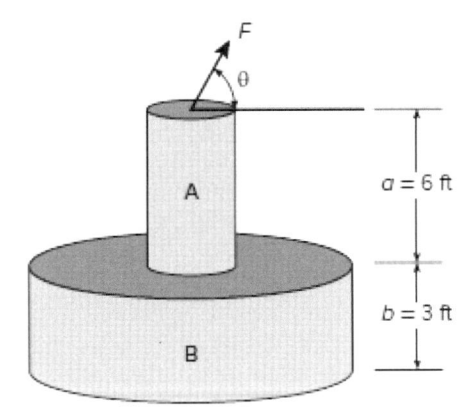

A. 0.25 in
B. 0.5 in
C. 1.75 in
D. 2.5 in

17. In the system below, what is the ratio of the lengths of cable 1 to cable 2 that keeps the rigid, weightless rod AB in a horizontal position? Assume that $E_1 = E_2$, $F = 800$ kN, $x = 1$ m, $A_1 = 2,600$ mm², $A_2 = 1,300$ mm², and $AB = L = 4$ m.

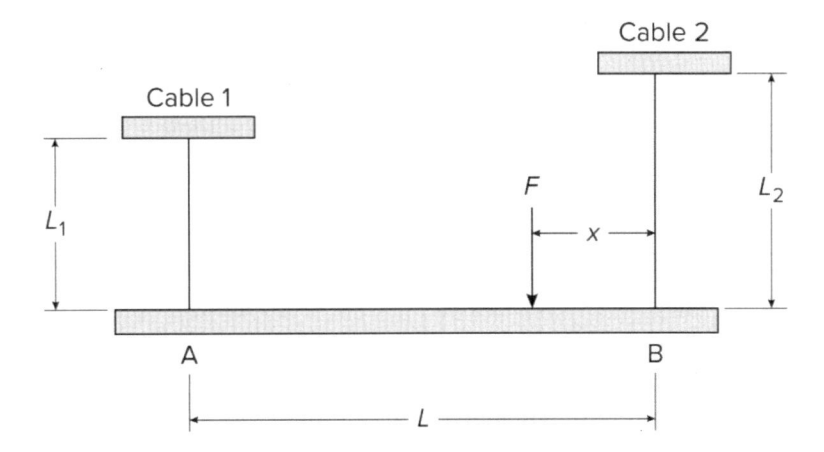

A. 0.25
B. 0.5
C. 1
D. 6

18. A 4-m cantilever beam is loaded with a distributed load of 1 kN/m. If the beam has a rectangular cross section that is 10 cm wide × 25 cm deep, what is the maximum shearing stress at the support?

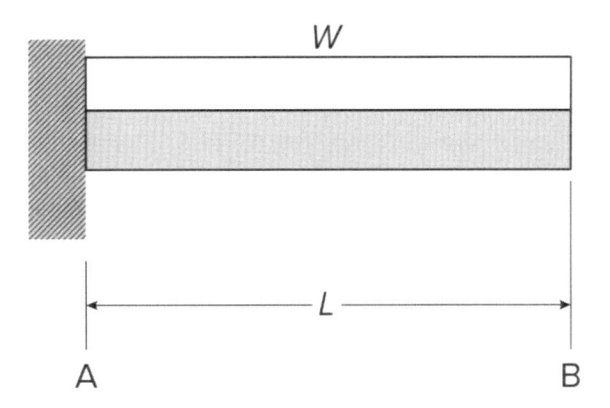

[Figure: cantilever beam fixed at A, free at B, length L, distributed load W on top]

- A. τ_{max} = 240 kPa
- B. τ_{max} = 260 kPa
- C. τ_{max} = 270 kPa
- D. τ_{max} = 380 kPa

19. Two 18-in diameter pipes carry water into a pipe junction, and a single 30-in diameter pipe carries water out of the junction. The velocities in the two 18-in pipes are 2.5 ft/s and 3.8 ft/s. What is the velocity in the 30-in pipe?

- A. 1.98 ft/s
- B. 2.27 ft/s
- C. 3.15 ft/s
- D. 4.27 ft/s

20. A pipe system with a corrugated metal pipe (n = 0.024) and a slope of 0.02 ft/ft is being designed. The desired flow is 200 ft³/s. What is the smallest diameter pipe required to carry this flow?

- A. 1.0 ft
- B. 3.4 ft
- C. 5.0 ft
- D. 9.6 ft

PRACTICE EXAM

21. A pressurized water main is 400 ft long with a 24-in diameter, a Hazen-Williams friction factor C of 110, and a head loss of 2.4 ft. What is the average velocity in this pipe?

- A. 4.12 ft/s
- B. 5.91 ft/s
- C. 7.43 ft/s
- D. 9.12 ft/s

22. A sanitary sewer 15 inches in diameter carries an open channel flow. The downstream end of the pipe is open, and the water flow from the pipe forms a waterfall. At the end of the pipe, the depth is equal to half of the pipe diameter. Compute the discharge Q in the pipe.

- A. 0.62 ft³/s
- B. 1.23 ft³/s
- C. 2.41 ft³/s
- D. 24.1 ft³/s

23. Water is flowing through 55 ft of 8-in pipe at a rate of 250 gpm. The water also travels through three check valves (K = 0.3) and two gate valves (K = 0.15). The friction losses through the pipe are known to be 0.65 ft per 100 ft of pipe. What are the total losses in the system?

- A. 0.21 ft
- B. 0.32 ft
- C. 0.41 ft
- D. 0.80 ft

24. Runoff from a 3-acre site is to be drained by a channel. The time of concentration for this site is 40 min. The site has a runoff coefficient C of 0.2. Rainfall quantities to be used for design are 0.5 in for a 20-min storm, 0.7 in for a 40-min storm, and 0.9 in for a 60-min storm. For what discharge should this channel be designed?

- A. 0.36 ft³/s
- B. 0.63 ft³/s
- C. 1.04 ft³/s
- D. 4.30 ft³/s

25. The centerline of a four-lane roadway is a horizontal circular curve as shown below. It is known that the PC station is sta 8+60, the curve radius R is 2,480 ft, and the intersection angle I is 70°. What is most nearly the PT station?

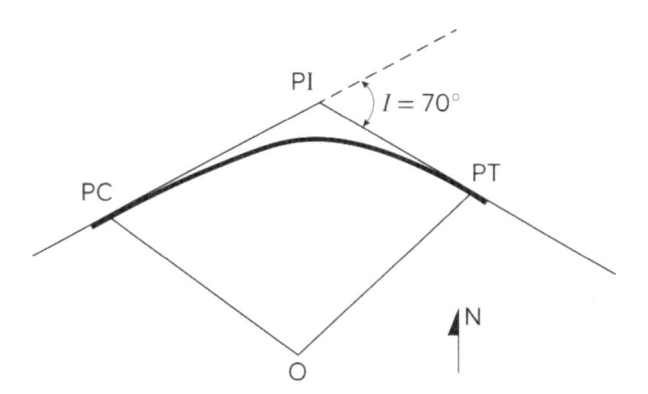

A. sta 38+15
B. sta 38+60
C. sta 39+89
D. sta 38+90

26. Assume that a speed-density study has resulted in the following calibrated relationship between space mean speed (v, mph) and density (k, veh/mi/ln): $v = 55 - 0.45k$. Determine the jam density and free-flow speed.

A. 0 veh/mi/ln, 55 mph
B. 120 veh/mi/ln, 50 mph
C. 122.2 veh/mi/ln, 0 mph
D. 122.2 veh/mi/ln, 55 mph

27. In horizontal curve formulas, what is tangent distance?

A. The distance from the PC to PI or from the PI to PT
B. The distance from the PI to the middle point of the curve
C. The distance along the line joining the PC and the PT
D. The distance from the middle point of the curve to the middle of the chord joining the PC and PT

PRACTICE EXAM

28. A horizontal curve is designed with an intersection angle of 13°21′55″. The degree of curvature (D) is 7.74°. Determine the tangent distance (T) to the nearest ft.

- A. 87 ft
- B. 118 ft
- C. 542 ft
- D. 740 ft

29. Increasing the water content in a concrete batch mix will result in (select all that apply):

- A. decreased strength.
- B. increased slump.
- C. increased sulfate resistance.
- D. increased strength.
- E. decreased slump.

30. A soil fill sample has a weight of 62 lb, a total volume of 864 in³, and a water content of 15%. Determine the percent relative compaction of the sample if the maximum dry unit weight is 115 lb/ft³.

- A. 93.8%
- B. 95.2%
- C. 102%
- D. 127%

31. Based on the boring log shown below, what is most nearly the buoyant (submerged) unit weight of soil at a depth of 15 ft? Assume that the soil below the water table is saturated.

DEPTH (ft)	N VALUE	UNIFIED SOIL CLASSIFICATION	DEPTH TO WATER (ft) 5.0 — MOISTURE CONTENT (%)	DEPTH TO WATER (ft) 5.0 — DRY DENSITY (lb/ft^3)
0	10	SM	10%	105
5				
6	7	CL	27%	90
10				
12				
15	8	CL	25%	95
19				
20	9	CL	21%	100

A. 33 lb/ft^3
B. 56 lb/ft^3
C. 95 lb/ft^3
D. 119 lb/ft^3

32. A soil has a void ratio of 0.7 and a water content of 22%. What is the total unit weight of the soil if the soil solids have a specific gravity of 2.7?

A. 102 lb/ft^3
B. 112 lb/ft^3
C. 121 lb/ft^3
D. 210 lb/ft^3

33. The construction of a new concrete pad has been commissioned to replace an existing one that has begun to deteriorate. The proposed slab is located in a coastal environment where there is a high presence of sulfates. In addition, the owner has mentioned that they will impose liquidated damages in the event the concrete does not meet expectations. Which type of cement should be specified for the batch mix design?

A. Type I
B. Type II
C. Type III
D. Type V

34. What is the plastic section modulus ratio of hot-rolled steel sheet pile PZ 22 to PZ 35?

- A. 0.37
- B. 0.38
- C. 0.63
- D. 2.62

35. After the placement of a reinforced concrete wall in a commercial building, the owner suspects internal defects in the structural element due to the low slump of concrete (2 in) and wants to examine this possibility. No damage to the structure is allowed by the engineer. Which test method(s) can be used to identify the potential internal defects? Select all that apply.

- A. Impact-echo
- B. Sounding
- C. Infrared thermography
- D. GPR
- E. Impulse response

36. The Department of Transportation is looking to construct a proposed earthen road using borrow soil. The road will be shaped in the form of a trapezoidal cross section. The top width of the roadway will be 50 ft. The height of the roadway from existing grade is 4.5 ft. The roadway should be constructed with a minimum slope of 3:1 (H:V). The length of the new road will be 5 miles. A sample of the imported soil is tested and yields a shrinkage percentage of 20% and a swell percentage of 35%. What is the minimum bank volume required to construct the roadway?

- A. 279,400 yd³
- B. 335,280 yd³
- C. 349,250 yd³
- D. 429,846 yd³

37. Given the earthwork summary at the following stations, determine the net quantity and resultant condition between sta 1+25 and sta 3+50 using the average end-area method.

STATION	FILL (ft²)	CUT (ft²)
1+25	110	20
3+50	50	230
4+50	30	42

    A. 375 yd³ (cut)
    B. 666.7 yd³ (fill)
    C. 1,041.7 yd³ (cut)
    D. 1,708.3 yd³ (fill)

38. A superintendent has been asked to compute the safety incidence rate for the last calendar year for insurance reporting purposes. From January to November, there have been a total of six serious injuries and three moderate illnesses due to exposures. From November to December, there was only one serious injury. There were 135 active employees during the reporting period, each working 56 hours per week for 50 weeks. The computed incidence rate (IR) is most nearly:

    A. 3.70
    B. 4.76
    C. 5.29
    D. 6.24

39. A recent geotechnical investigation revealed that the layered soil adjacent to the existing building can be classified as type C over type B according to OHSA regulation 1926 subpart P, "Excavations." An excavation is planned to upgrade subsurface utility structures. A 3-ft undisturbed perimeter is required from the face of the building, the total depth of excavation is 13 ft, and the top layer of soil is 6 ft deep. According to OSHA, the minimum horizontal distance from the face of the building to the toe of the slope is most nearly:

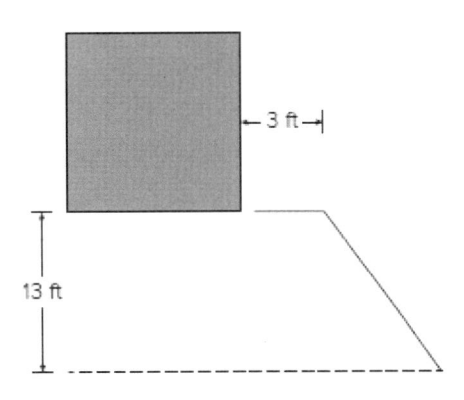

- A. 16 ft
- B. 19 ft
- C. 19.5 ft
- D. 22 ft

40. Based on the information provided in the figure below, the difference in elevation from the first benchmark (BM$_1$) to the second benchmark (BM$_2$) is most nearly:

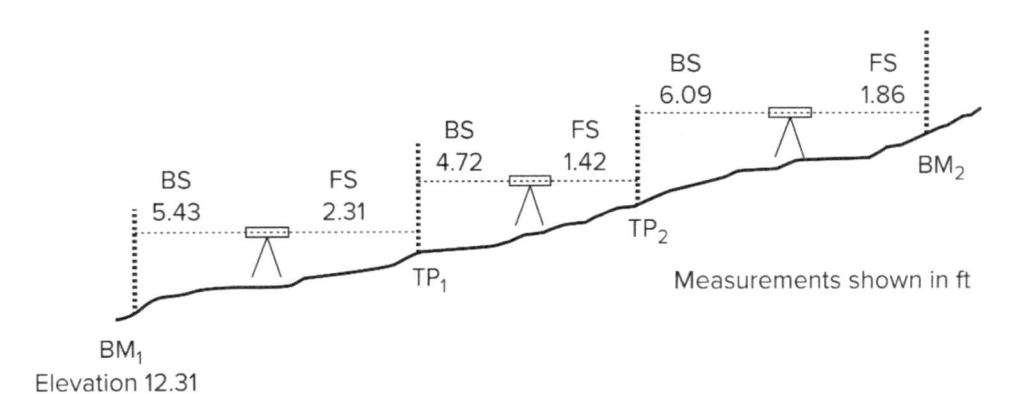

- A. 1.66 ft
- B. 10.65 ft
- C. 22.96 ft
- D. 28.55 ft

41. For a four-leg intersection, compared to conventional intersections, the number of potential conflict points of roundabouts is:

A. less.
B. more.
C. the same.
D. undetermined.

42. Suppose that an individual can travel to work by driving alone (A), carpooling (C), or riding a bus (B). The utility function for the transportation modes is $U = -T - 5C/Y$, where the values are provided in the table. Use the given logit model to determine the probability that an individual would select carpooling to work.

$$P_i = \frac{e^{U_i}}{\sum_{j=1}^{n} e^{U_j}}$$

P_i = probability that users with utility values U_i will select mode i
U_i = utility of mode i
n = number of modes

	DOOR-TO-DOOR TRAVEL TIME, T (hr)	TRAVEL COST, C ($)	ANNUAL INCOME, Y ($1,000/yr)
Driving alone	0.50	3.00	50
Carpooling	0.75	1.20	50
Bus	1.00	0.80	50

A. 0.153
B. 0.251
C. 0.347
D. 0.632

43. Which statements are true regarding signalized intersections? Choose all that apply.

A. Signalized intersections are controlled by signals operating in only two phases.
B. The capacity (in veh/hr) of a signalized intersection is calculated for each lane group.
C. Each phase of a signal consists of three intervals: red, yellow, and green.
D. Level of service for signalized intersections is defined in terms of control delay in the intersection.

PRACTICE EXAM

44. Four vehicles travel on a 300-ft section of road with their travel times and speeds recorded in the table. The space mean speed is most nearly:

SPEED (ft/s)	TIME (s)
30	10.0
40	7.5
50	6.0
60	5.0

- A. 40 ft/s
- B. 42 ft/s
- C. 43 ft/s
- D. 50 ft/s

45. Scott has three options to get to work: bus, car, or train. The average utilities of the three modes are -0.02, -0.3, and -0.01, respectively. Use the given logit model to determine the probability that Scott will take the train to get to work.

$$P_i = \frac{e^{U_i}}{\sum_{j=1}^{n} e^{U_i}}$$

P_i = probably that users with utility values U_i will select mode i
U_i = utility of mode i
n = number of modes

- A. 0.167
- B. 0.365
- C. 0.333
- D. 0.412

46. For what purpose is the level of service (LOS) for pedestrians used? Choose all that apply.

- A. To evaluate the performance of a sidewalk
- B. To determine the need for a bike lane
- C. To remove street furniture
- D. To change the width of a sidewalk

PRACTICE EXAM

47. Which of the following highway crossings is appropriate for the intersection of two freeways with turning traffic between them?

- A. At-grade intersections
- B. Grade separation without ramps
- C. Interchanges
- D. All of the above

48. There are three rear-end, six left-turn, and seven right-angle crashes within one year at a roadway intersection. The average daily traffic (ADT) entering the intersection is 1,000. What is the left-turn crash rate per million entering vehicles at this intersection?

- A. 12.6
- B. 16.4
- C. 17.1
- D. 19.7

49. The following information is given for a signalized intersection:

Standard vehicle length = 20 ft
Average vehicle approach speed = 37.8 ft/s
Length of red clearance interval = 1.8 s
The width of the intersection is most nearly:

- A. 24 ft
- B. 36 ft
- C. 42 ft
- D. 48 ft

50. Find the proportion of vehicles arriving to an intersection during green, assuming a cycle length (C) of 60 s, an effective green time (G) of 29 s, and a platoon ratio of 1.

- A. 0.25
- B. 0.36
- C. 0.48
- D. 0.60

PRACTICE EXAM

51. If the average daily traffic (ADT) entering one roadway intersection is 300, what is the annual traffic entering the intersection?

    A. 10,950
    B. 109,500
    C. 136,520
    D. 142,300

52. Point X is a POC located at sta 16+50.00. Point B is the curve PT and is located at sta 20+00.00. The degree of curvature is 6.75°. Using the information provided, determine the coordinates of point X.

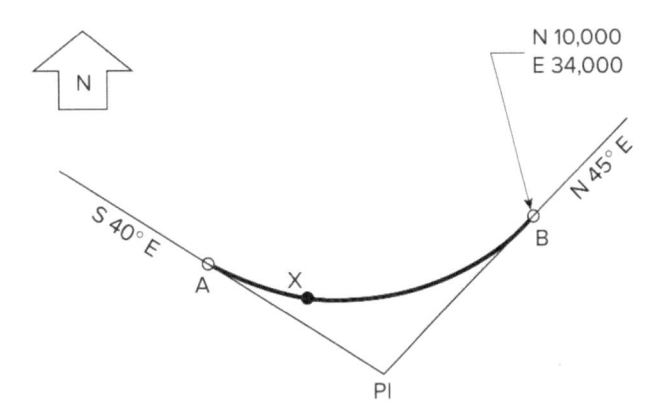

    A. N 4,602.34   E 30,234.78
    B. N 9,549.12   E 31,456.11
    C. N 9,809.77   E 33,709.16
    D. N 9,949.00   E 33,709.16

53. A horizontal curve has a deflection angle of 40°30′, degree of curvature of 9°30′, and a point of intersection located at sta 21+00. The station of point of tangent for this curve is most nearly:

    A. 21+00.65
    B. 22+00
    C. 23+03.78
    D. 23+10

PRACTICE EXAM

54. There is an obstruction on a curve. The radius of the curve is 45 ft, and the sight distance is 40 ft. What is the horizontal sightline offset (HSO) for this curve?

- A. 2.5 ft
- B. 3.0 ft
- C. 4.0 ft
- D. 4.4 ft

55. An existing horizontal curve has a radius of 85 m that restricts the maximum speed on this section of road to only 60 km/h, which represents 60% of the design speed. Highway officials want to improve the road to eliminate this speed reduction. Assume a coefficient of side friction of 0.15 and a superelevation rate of 8%. What is the new radius of curvature?

- A. 321 m
- B. 342 m
- C. 356 m
- D. 385 m

56. What is the length of a vertical curve with S = 500 ft, differential slope of 6%, standard conditions, and $L = S + 195.1$ ft based on a crest vertical curve and stopping sight distance?

- A. 456 ft
- B. 623 ft
- C. 695 ft
- D. 786 ft

57. The stopping sight distance (SSD) of a vehicle on a level roadway traveling at a design speed of 50 mph with a deceleration rate of 4.0 ft/s² is 855 ft. Assuming the same brake reaction time, what will the SSD become if the vehicle travels on an upgrade with a 3% positive slope?

- A. 725 ft
- B. 855 ft
- C. 975 ft
- D. 1,043 ft

PRACTICE EXAM

58. You are driving with a structure overhead and want to see past the object on the sag curve, the end of which is within your line of sight. The difference in grade is 3%, the sight distance is 1,500 ft, and the overhead structure is 15 ft. Your eyes are 3 ft from the road and an object ahead is 5 ft from the road. Find the length of the curve.

- A. 32 ft
- B. 57 ft
- C. 59 ft
- D. 67 ft

59. A driver is traveling down a two-lane highway at 75 mph when he sees a sign warning of a roadway obstruction 1,000 ft ahead. The highway is on a downhill grade of 3.2%. Assume friction is 0.27. Considering his reaction time of 1.5 s, the distance he travels before the vehicle comes to a stop is most nearly:

- A. 953 ft
- B. 979 ft
- C. 1,026 ft
- D. 1,041 ft

60. According to AASHTO, an auxiliary lane should be provided for speed-change lanes when the distance between successive noses is less than:

- A. 1,500 ft
- B. 1,600 ft
- C. 1,800 ft
- D. 2,000 ft

61. A five-lane highway is being redesigned in an urban downtown area. The engineer is adding raised curb sections with landscapes. Refuge islands will be installed at the intersections. What is the required minimum width of the refuge island to accommodate the pedestrians according to AASHTO?

- A. 4 ft
- B. 6 ft
- C. 8 ft
- D. 10 ft

PRACTICE EXAM

62. Lisa is test driving a new sports car along a minor road and has arrived at an intersection with stop control to a major four-lane highway. She needs to make a left turn from stop on the minor road and sees a vehicle approaching from the left. Determine the length of leg of the sight triangle along the major road needed to make a left turn given the speeds on the major and minor road are 50 mph and 35 mph, respectively. The minor road grade is 4%.

  A. 392 ft
  B. 447 ft
  C. 551 ft
  D. 647 ft

63. A taper-type deceleration lane is being constructed for a new exit ramp along a freeway with a design speed of 60 mph. The exit curve design speed is 40 mph. Determine the minimum deceleration length of the exit curve given a 4% downgrade.

  A. 300 ft
  B. 350 ft
  C. 360 ft
  D. 420 ft

64. A 4,000-lb vehicle is travelling at a speed of 76 ft/s. Find the mass of sand needed to reduce the velocity of the vehicle to 70 ft/s.

  A. 332 lb
  B. 343 lb
  C. 354 lb
  D. 361 lb

65. Find the length of need (X) for a barrier installed parallel to the road. Consider the following characteristics:

a. Lateral extent of area of concern (L_A) = 25 ft
b. Lateral distance from edge of traveled way (L_2) = 15 ft
c. Runout length (L_R) = 300 ft

  A. 120 ft
  B. 130 ft
  C. 220 ft
  D. 300 ft

66. Which of the following options represent concrete barriers? Select all that apply.

- A. Trinity T-39
- B. New Jersey barrier
- C. Three-beam guardrail
- D. U-GUARD 31
- E. Low profile barrier
- F. Constant slope barrier

67. What is the maximum curb ramp grade for accessibility?

- A. 8.00%
- B. 8.33%
- C. 8.66%
- D. 9.00%

68. The following diagram and information are given for an intersection near a grade crossing controlled by a stop sign on a minor street.

Given:
Distance (*D*) = 95 ft
Major street information:
Major street traffic volume (both approaches) = 250 vph
Minor street information:
Minor street traffic volume (one approach) = 150 vph
Rail traffic/day = 1
2% high-occupancy buses
20% tractor trailer trucks

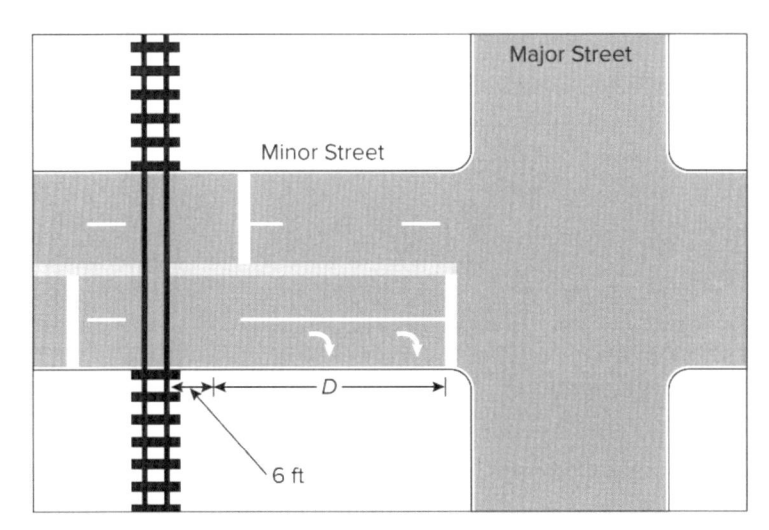

Source: Federal Highway Administration (FHWA). 2009. *Manual on Uniform Traffic Control Devices (MUTCD)*. Washington, DC: US Department of Transportation.

A traffic signal is warranted in this intersection because the adjusted minor street traffic volume is _____.

Round to the nearest whole number and fill in the blank.

69. Four intersections on a busy city street have been studied for 8 months to determine the accident rate. The city only has enough money to improve two of the intersections. Based on the information, which two intersections will the city most likely improve?

INTERSECTION	NUMBER OF ACCIDENTS	ADT
1	5	8,000
2	25	26,500
3	34	42,300
4	16	19,000

A. 1 and 3
B. 2 and 3
C. 2 and 4
D. 3 and 4

70. At a signalized intersection, the signal cycle length (C) = 90 s. For approach B, the green time (G) = 56 s, the yellow plus all-red time (Y) = 4 s, the start-up lost time (L_1) = 2 s, the clearance lost time (L_2) = 1 s, and the maximum number of vehicles on approach B that can traverse the intersection during green time is 1,600 vph per lane. Determine the capacity of approach B at the signalized intersection.

A. 783 veh/(hr · lane)
B. 1,220 veh/(hr · lane)
C. 1,570 veh/(hr · lane)
D. 1,049 veh/(hr · lane)

71. According to the *MUTCD*, stripes on barricades should be alternating orange and white stripes at what angle in the direction road users are to pass?

A. 30 degrees
B. 45 degrees
C. 60 degrees
D. 75 degrees

PRACTICE EXAM

72. A local utility company is planning to install conduit in the center of an intersection in downtown West Haven. The intersection must be closed and requires a temporary traffic control plan to perform the work. The posted speed limit of the facility is 30 mph, width of the lane is 12 ft, and the length of lane closure is 180 ft. Determine the minimum length of the taper required to channelize traffic away from the center of the intersection.

- A. 90 ft
- B. 100 ft
- C. 180 ft
- D. 200 ft

73. Edge line markings should be placed on paved streets with all of the following characteristics except:

- A. freeways.
- B. expressways.
- C. on paved streets with traveled ways delineated by curbs, parking, or other markings.
- D. collectors with a traveled way of ≥ 20 ft and an ADT of ≥ 3,000 vpd.

74. An asphalt pavement consists of a 6-in thick dense-graded asphalt concrete (AC) layer and an aggregate base (AB) layer over a subgrade. The layer coefficients are 0.40 and 0.16 for the AC and AB layers, respectively. The required structural number (SN) for this pavement structure is 4.0. Determine the design thickness (D_2) of the AB layer.

- A. 4 in
- B. 6 in
- C. 8 in
- D. 10 in

75. In 2019, a six-lane rural interstate has a daily truck count of 3,500 vpd. Considering: truck factor of 0.52, directional factor of 0.50, lane distribution factor of 0.70, and annual growth rate of 5%; the traffic loading difference, in equivalent single axle loads (ESALs), between 2019 and 2023 is most nearly:

- A. 232,500
- B. 485,520
- C. 769,615
- D. 835,000

76. The following data were obtained by monitoring traffic and determining the truck axle loads on a highway.

TRAFFIC DATA	
Initial ADT	18,000 vpd
Growth rate	2.5%/yr
Design period	20 yr
Fraction of truck traffic	0.15

AXLE LOAD DATA	
Axle load (lbf)	Number of axles
SINGLE AXLES	
4,000	100
8,000	950
12,000	1,870
22,000	60
26,000	20
DOUBLE AXLES	
6,000	510
10,000	620
32,000	100
TRIPLE AXLES	
10,000	300
18,000	400
44,000	60

Assuming a structural number of 2 and a terminal serviceability value (p_t) of 2.5, determine the total daily 18-kip equivalent axle load (ESAL) applications.

    A. 600
    B. 770
    C. 806
    D. 930

PRACTICE EXAM

77. The results of a recent soils survey for a proposed highway are shown below. If the project specifies that soils stabilization is needed for AASHTO classified soils A-6 and A-7, which locations will require stabilization?

LOCATION	LIQUID LIMIT, %	PLASTIC LIMIT, %	PASSING N200, %
1	23	12	8
2	49	15	55
3	55	19	36

A. 1 only
B. 1, 2
C. 1, 2, 3
D. 2, 3

78. If a culvert is flowing partially full, which of the following types of flow is not possible?

A. Critical
B. Subcritical
C. Pressurized
D. Supercritical

79. Determine the travel time for 400 m of stormwater flowing through a pipe with a slope of 0.005. Use Manning's equation and a roughness coefficient of $n = 0.011$ and $R_h = 0.2$ for the flow.

A. 2.0 min
B. 2.7 min
C. 3.0 min
D. 3.2 min

80. The table shows the details for three highway improvement proposals under review by the local DOT. If the DOT suggests a 2% discount rate, the difference between the two most economical options is most nearly:

ALTERNATIVE	INITIAL COST	ANNUAL MAINTENANCE	REHAB AT YEAR 10
A	$1,000,000	$20,000	$300,000
B	$500,000	$50,000	$500,000
C	$1,500,000	$10,000	$0

A. $100,000
B. $200,000
C. $1,400,000
D. $1,500,000

SOLUTIONS

QUICK SOLUTIONS REFERENCE

Question #	Answer
1.	C
2.	A
3.	355
4.	B
5.	51
6.	C
7.	C
8.	C
9.	D
10.	B
11.	A
12.	A
13.	D
14.	C
15.	C
16.	A
17.	D
18.	A
19.	B
20.	C
21.	B
22.	C
23.	C
24.	B
25.	D
26.	D
27.	A
28.	A
29.	A, B
30.	A
31.	B
32.	C
33.	D
34.	B
35.	A, E
36.	C
37.	A
38.	C
39.	B
40.	B

Question #	Answer
41.	A
42.	C
43.	B, C, D
44.	B
45.	B
46.	A, C, D
47.	C
48.	B
49.	D
50.	C
51.	B
52.	C
53.	C
54.	D
55.	B
56.	C
57.	A
58.	D
59.	A
60.	A
61.	B
62.	D
63.	D
64.	B
65.	A
66.	B, E, F
67.	B
68.	148
69.	C
70.	D
71.	B
72.	A
73.	C
74.	D
75.	C
76.	C
77.	D
78.	C
79.	C
80.	A

SOLUTIONS

1. First, calculate the length of the channel wall:

$$\text{Hyp} = \left\{\sqrt{(42 \text{ ft})^2 + \left[\frac{1}{4}(42 \text{ ft})\right]^2}\right\} = 43.29 \text{ ft}$$

Calculate the volume of concrete given the 7.5% waste:

$$V_{\text{Concrete}} = \left(\frac{8 \text{ in}}{12 \text{ in}}\right)\{[650 \text{ ft}][30 \text{ ft} + 2(43.29 \text{ ft})]\} = 50{,}518 \text{ ft}^3$$

$$+ \text{ Waste } = 1.075(V_{\text{Concrete}}) = 54{,}306.9 \text{ ft}^3$$

Convert to cubic yards:

$$V_{\text{Concrete}} = (54{,}309.3 \text{ ft}^3)\left(\frac{1 \text{ yd}^3}{27 \text{ ft}^3}\right)$$

$$V_{\text{Concrete}} = 2{,}011.5 \text{ yd}^3$$

Reference: NCEES *PE Civil Reference Handbook* > Construction > Estimating Quantities and Costs
Answer: C

2. Perform the forward pass to determine the early dates (ES and EF). Then, perform the backward pass to determine the late dates (LS and LF). Once both the early and late dates are determined, compute the total float by employing the following equation: TF = LS − ES or LF − EF.

	ES	LS	EF	LF	FLOAT
1	0	0	3	3	0
2	3	7	13	17	4
3	3	3	9	9	0
4	9	9	17	17	0
5	17	17	21	21	0

Alternatively, recognize that after the forward pass, activity 5 is present on the critical path, and the total and free float of critical activities are always equal to zero. Therefore, the total float of task 5 is zero.

Reference: NCEES *PE Civil Reference Handbook* > Construction > Scheduling > Critical Path Method (CPM) Network Analysis
Answer: A

SOLUTIONS

3. Determine the cost of the entire building:

$$(100{,}000 \text{ ft}^2)\left(\frac{\$360}{90 \text{ ft}^2}\right) = \$400{,}000$$

Determine the budget remaining after the cost of the building and how many parking spots can be built with that amount:

$\$2{,}000{,}000 - \$400{,}000 = \$1{,}600{,}000$

$$\text{Number of parking spots} = \frac{\$1{,}600{,}000}{\$4{,}500/\text{spot}} = 355.55 = 355 \text{ spots}$$

Reference: NCEES *PE Civil Reference Handbook* > Construction > Estimating Quantities and Costs

Answer: 355

4. The following relationships are true for any activity:
TF = LF − EF

EF = ES + D

Substitute:
$TF_C = LF_C - (ES_C + D_C) = LF_C - ES_C - D_C$

Reference: NCEES *PE Civil Reference Handbook* > Construction > Scheduling > Critical Path Method (CPM) Network Analysis

Answer: B

SOLUTIONS

5. Determine the minimum daily production rate to complete the scope (or the amount of material that must be hauled in a given day to meet the schedule of 15 days):

$$\frac{182{,}212 \text{ yd}^3/\text{job}}{15 \text{ days/job}} = 12{,}147.47 \text{ yd}^3/\text{day}$$

Determine how much material a single truck can haul in a given 8-hr day, recognizing that a single truck can transport 20 yd³ per trip (cycle) over the span of 40 min (0.667 hr):

$$(20 \text{ yd}^3/\text{cycle})\left(\frac{8 \text{ hr/day}}{0.667 \text{ hr/cycle}}\right) = 240 \text{ yd}^3/\text{day (per truck)}$$

Determine how many trucks are required to support a production of 12,147.47 yd³ if every truck can haul 240 yd³ per day:

$$\frac{12{,}147.47 \text{ yd}^3/\text{day}}{240 \text{ yd}^3/\text{day (per truck)}} = 50.61 \text{ trucks} \approx 51 \text{ trucks}$$

Reference: NCEES *PE Civil Reference Handbook* > Construction Operations and Methods > Production Rate for Loading and Hauling Earthwork

Answer: 51

SOLUTIONS

6. Moment of force (or torque) is equal to the force multiplied by the perpendicular distance to the point of inquiry:
$$M = Fd_\perp = 750 \text{ in} \cdot \text{lb}$$

Solve for the perpendicular distance assuming a maximum force of 50 lb can be applied:
$$\frac{M}{F} = d_\perp = \frac{750 \text{ in} \cdot \text{lb}}{50 \text{ lb}} = 15 \text{ in}$$

Recognize that the 15 in will be composed of 2 distances: the distance from the center of the bolt to the offset, and the leg of the triangle (with a hypotenuse of *L*).
15 in – 1.75 in = 13.25 in

13.25 is the perpendicular distance from the bolt head to the application of force, *P*.

Solve for distance *L* using the angle:
$$\frac{13.25 \text{ in}}{\cos 30°} = 15.3 \text{ in}$$

Reference: NCEES *PE Civil Reference Handbook* > Statics
Answer: C

SOLUTIONS

7. Recognize that the opposing force components are attributed from the wind and brace resistance, respectively.

The maximum wind load is a function of the tributary area, which is determined as:
A_{Trib} = (height of the wall)(center-to-center brace spacing)
A_{Trib} = 12 ft(7 ft) = 84 ft²

Solve for the component of the maximum brace load:
$$F = \frac{F_x}{\cos \theta} = 4{,}158 \text{ lb}; \quad F_x = 4{,}158 \text{ lb}(\cos 45°) = 2{,}940.15 \text{ lb}$$

Solve for the sum of moments at point A to determine the opposing point load (P) generated from W_{Load} to F_x (the resultant of the wind load (P) is applied at the middle of the wall height):
$$\Sigma M_A = \left(\frac{12}{2}P\right) - [(6 \text{ ft})(2{,}940.15 \text{ lb})] = 0; \quad P = 2{,}940.15 \text{ lb}$$

Solve for maximum W_{Load} given A_{Trib}:
$$P = W_{Load} A_{Trib} = 2{,}940.15 \text{ lb}; \quad \frac{2{,}940.15 \text{ lb}}{84 \text{ ft}^2} = 35 \text{ lb/ft}^2$$

Reference: *NCEES PE Civil Reference Handbook* > Statics
Answer: C

8. The diagram shows that there are three layers of soil with unit weights of 16 kN/m³, 17 kN/m³, and 18 kN/m³.

The total stress at point A is calculated as:
$\sigma = \gamma_1 H_1 + \gamma_2 H_2 + \gamma_3 H_3$
$\sigma = (16 \text{ kN/m}^3)H_1 + (17 \text{ kN/m}^3)(3 \text{ m}) + (18 \text{ kN/m}^3)(2 \text{ m})$
$\sigma = (87 + 16H_1)\text{kPa}$

The pore water pressure is given as:
$u = \gamma_w(H_2 + H_3) = (9.8 \text{ kN/m}^3)(3 \text{ m} + 2 \text{ m}) = 49 \text{ kPa}$

Effective pressure at point A is calculated as:
$\sigma' = \sigma - u = (87 + 16H_1)\text{kPa} - 49 \text{ kPa} = 90 \text{ kPa}$

The above equation shows:
$H_1 = 3.25 \text{ m} \approx 3.3 \text{ m}$

Reference: NCEES *PE Civil Reference Handbook* > Geotechnical > Effective and Total Stresses
Answer: C

SOLUTIONS

9. It is known that:

Water content, $\omega = 100 \dfrac{W}{W_S}$

Specific gravity of soil solids, $G_S = \dfrac{W_S}{V_S \gamma_w}$

Void ratio, $e = \dfrac{V_V}{V_S}$

Degree of saturation, $S = 100 \dfrac{V_W}{V_V}$

Water volume, $V_W = \dfrac{W_W}{\gamma_w}$

Therefore:

$Se = 100 \left(\dfrac{V_W}{V_V}\right)\left(\dfrac{V_V}{V_S}\right) = 100 \dfrac{V_W}{V_S}$ and $G_S\omega = 100 \left(\dfrac{W_S}{V_S \gamma_w}\right)\left(\dfrac{W_W}{W_S}\right) = 100 \dfrac{V_W}{V_S}$

Which means that: $Se = G_S\omega$

The void ratio of the soil above the water level is calculated as:

$e = \dfrac{G_S \omega}{S} = \dfrac{2.7(20\%)}{0.5} = 1.08$

The unit weight of soil above the water level is calculated as:

$\gamma_A = \gamma_w \left[\dfrac{G_S + Se}{1 + e}\right] = 9.8 \text{ kN/m}^3 \left[\dfrac{2.7 + 0.5(1.08)}{1 + 1.08}\right] \cong 15.27 \text{ kN/m}^3$

The void ratio of the soil below the water level is calculated as:

$e = \dfrac{G_S \omega}{S} = \dfrac{2.7(35\%)}{1.0} = 0.945$

The unit weight of soil below the water level is calculated as:

$\gamma_B = \gamma_w \left(\dfrac{G_S + e}{1 + e}\right) = 9.8 \text{ kN/m}^3 \left(\dfrac{2.7 + 0.945}{1 + 0.945}\right) \cong 18.37 \text{ kN/m}^3$

Effective stress is equal to the total stress minus pore water pressure:
$\sigma' = \sigma - u$.

The total stress is calculated as: $\sigma = \gamma_A H_1 + \gamma_B H_2$
$\sigma = (15.27 \text{ kN/m}^3)(3 \text{ m}) + (18.37 \text{ kN/m}^3)(5 \text{ m}) = 137.66 \text{ kPa}$

The pore water pressure is calculated as:
$u = \gamma_w H_2 = (9.8 \text{ kN/m}^3)(5 \text{ m}) = 49 \text{ kPa}$

The effective stress is calculated as:
$\sigma' = \sigma - u = 137.66 \text{ kPa} - 49 \text{ kPa} = 88.66 \text{ kPa}$

Reference: NCEES *PE Civil Reference Handbook* > Effective and Total Stresses & Material Test Methods

Answer: D

SOLUTIONS

10. Though a soil friction angle is not given, it is related to the slope of the failure plane, which can be used directly to compute the Rankine active earth pressure coefficient:

$K_a = \tan^2\left(45 - \frac{\phi}{2}\right) \rightarrow \alpha = 45 + \frac{\phi}{2} \rightarrow K_a = \tan^2(90 - \alpha) = \tan^2(30) = 0.33$

The horizontal stress at the base of the wall is:

$\sigma_h = K_a \gamma H$

$\sigma_h = (0.33)(115 \text{ lb/ft}^3)(15 \text{ ft})$

$\sigma_h = 569.25 \text{ lb/ft}^2$

The active lateral earth pressure resultant is then:

$R_a = \sigma_h H / 2$

$R_a = \dfrac{(569.25 \text{ lb/ft}^2)(15 \text{ ft})}{2}$

$R_a = 4{,}269.38 \text{ lb/ft} \approx 4{,}270 \text{ lb/ft}$

Reference: NCEES *PE Civil Reference Handbook* > Lateral Earth Pressures > Rankine Earth Coefficients

Answer: B

SOLUTIONS

11. Settlement for soil in the range of virgin compression is calculated as:

$$\Delta H = \frac{H_0}{1+e_0}\left[C_C \log \frac{p_0 + \Delta p}{p_0}\right]$$

In the equation above, all values are given in the question except the coefficient of consolidation (C_C) and the initial void ratio (e_0). Settlement is also referenced in units of feet.

First determine the initial void ratio:

$$e = \frac{n}{1-n} = \frac{0.5}{1-0.5} = 1$$

Based on the equation above, the C_C is calculated as:

$$\Delta H = \frac{H_0}{1+e_0}\left[C_C \log \frac{p_0 + \Delta p}{p_0}\right] = \frac{20 \text{ ft}}{1+1}\left[C_C \log \frac{(3{,}000 \text{ lb/ft}^2 + 2{,}000 \text{ lb/ft}^2)}{3{,}000 \text{ lb/ft}^2}\right]$$
$$= 12 \text{ in} = 1 \text{ ft}$$

Therefore:
$C_C \approx 0.45$

References: NCEES *PE Civil Reference Handbook* > Geotechnical > Material Test Methods > Weight-Volume Relationships
NCEES *PE Civil Reference Handbook* > Geotechnical > Consolidation
Answer A

12. Settlement for soil is calculated as:

$$\Delta H = \frac{H_0}{1+e_0}\left[C_R \log \frac{p_C}{p_0} + C_C \log \frac{p_0 + \Delta p}{p_C}\right]$$

The equation above shows everything except for the initial void ratio.
The void ratio is calculated as:

$$e = \frac{V_V}{V_S} = \frac{V_V}{V - V_V} = \frac{45 \text{ cm}^3}{100 \text{ cm}^3 - 45 \text{ cm}^3} = 0.8$$

The settlement of the soil is calculated as:

$$\Delta H = \frac{5 \text{ m}}{1+0.8}\left[(0.1)\log\frac{400 \text{ kN/m}^2}{300 \text{ kN/m}^2} + (0.6)\log\frac{500 \text{ kN/m}^2}{400 \text{ kN/m}^2}\right] = 19.6 \text{ cm}$$

Reference: NCEES *PE Civil Reference Handbook* > Geotechnical > Consolidation
Answer: A

SOLUTIONS

13. The diagram shows that there are two layers of soil. For the first layer, the unit weight is equal to:

$\gamma_1 = 103$ lb/ft³

For the second layer, γ_2, the unit weight is unknown.
The effective stress at point A is calculated as:

$\sigma = S + \gamma_1 H_1 + \gamma_2 H_2$

$\sigma = 630 \text{ lb/ft}^2 + (103 \text{ lb/ft}^3)(12 \text{ ft}) + \gamma_2 (15 \text{ ft}) = 3{,}800 \text{ lb/ft}^2$

$\sigma = 1{,}866 \text{ lb/ft}^2 + \gamma_2 (15 \text{ ft}) = 3{,}800 \text{ lb/ft}^2$

The above equation shows:
$\Gamma_2 = 128.9$ lb/ft³

Reference: NCEES *PE Civil Reference Handbook* > Geotechnical > Effective and Total Stresses
Answer: D

SOLUTIONS

14. Take moment about B to the find the reaction at A (the distributed load is converted to a point load acting at the center of b):

$\sum M_B = 0$

$R_A(L) - w(b)\left(\dfrac{b}{2}\right) = 0$

$R_A = \dfrac{w(b)\left(\dfrac{b}{2}\right)}{L} = \dfrac{(10 \text{ kpf})(8 \text{ ft})\left(\dfrac{8 \text{ ft}}{2}\right)}{20 \text{ ft}} = 16 \text{ kips}$

$\sum F_y = 0$

$R_A + R_B - w(b) = 0$

$R_B = w(b) - R_A = 10 \text{ kpf } (8 \text{ ft}) - 16 \text{ kips} = 64 \text{ kips}$

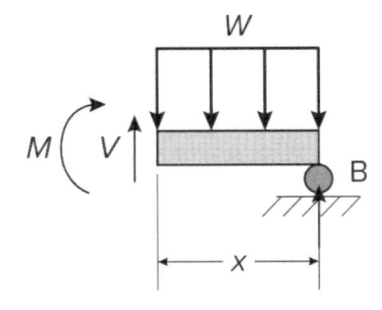

Cut a section at x from support B as seen in the beam free-body diagram above. The distance x from the right support with zero shear force is calculated as:

$\sum F_V = 0$

$R_B - w(x) + V = 0$

Set V equal to 0.

$R_B - w(x) = 0$

$64 \text{ kips} - 10\dfrac{\text{kips}}{\text{ft}(x)} = 0$

Therefore:

$x = \dfrac{64 \text{ kips}}{10 \text{ kpf}}$

$x = 6.4 \text{ ft}$

Reference: NCEES *PE Civil Reference Handbook* > Statics
Answer: C

SOLUTIONS

15. Based on geometry, the distributed load is converted to a concentrated load (W):

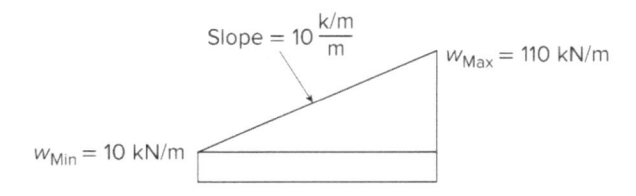

$W = (10 + 10x)$ kN/m from 0 to 10

The minimum intensity at the left support ($x = 0$ m):

$$w_{Min} = 10 \text{ kN/m} + 10\frac{\text{kN/m}}{\text{m}}(0 \text{ m}) = 10 \text{ kN/m}$$

The maximum intensity at point O (the right support, $x = 10$ m):

$$w_{Max} = 10 \text{ kN/m} + 10\frac{\text{kN/m}}{\text{m}}(10 \text{ m}) = 110 \text{ kN/m}$$

Find the resultant of the rectangular distributed load:

$$W_1 = w_{Min}L = 10\frac{\text{kN}}{\text{m}}(10 \text{ m}) = 100 \text{ kN}$$

Find the resultant of the triangular distributed load:

$$W_2 = \frac{(w_{Max} - w_{Min})L}{2} = \frac{(110 \text{ kN/m} - 10 \text{ kN/m})10 \text{ m}}{2} = 500 \text{ kN}$$

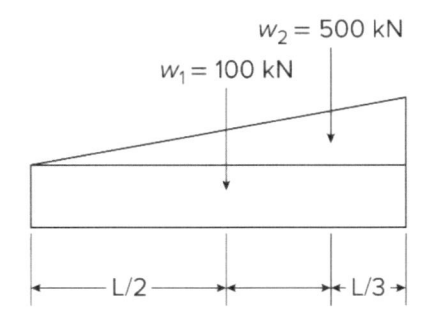

The shear force is calculated as

$\Sigma F = 0$

$R - W_1 - W_2 + V = 0$

$V = -R + W_1 + W_2 = -200 \text{ kN} + 100 \text{ kN} + 500 \text{ kN}$

$V = 400 \text{ kN}$

SOLUTIONS

The moment is calculated as:
$\Sigma M_O = 0$ (assume CW moment is positive)
$R(10 \text{ m}) - W_1 \left(\dfrac{10 \text{ m}}{2}\right) - W_2 \left(\dfrac{10 \text{ m}}{3}\right) + M = 0$

$M = (-200 \text{ kN})(10 \text{ m}) + (100 \text{ kN})(5 \text{ m}) + (500 \text{ kN})\left(\dfrac{10 \text{ m}}{3}\right)$

$M = -2{,}000 \text{ kN} \cdot \text{m} + 500 \text{ kN} \cdot \text{m} + 1{,}666.7 \text{ kN} \cdot \text{m}$

$M = 166.7 \text{ kN} \cdot \text{m}$

Reference: NCEES *PE Civil Reference Handbook* > Statics
Answer: C

16. Draw the following free-body diagram:

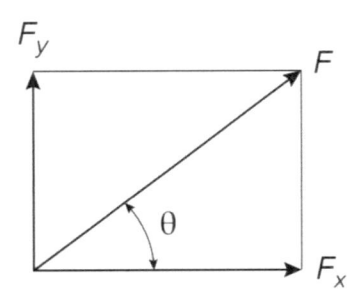

The force component that causes elongation is F_y:
$\Sigma F_y = 0$

$F_y = F \sin \theta = 800 \text{ kips} (\sin 30°) = 400 \text{ kips}$

Elongation is calculated as:
$\delta_{Total} = \delta_A + \delta_B = \dfrac{F_y L_1}{EA_A} + \dfrac{F_y L_2}{EA_B}$

$\delta_{Total} = \dfrac{(400 \text{ kips})(6 \text{ ft})(12 \text{ in/ft})}{(5 \text{ in}^2)(29{,}000 \text{ kips/in}^2)} + \dfrac{(400 \text{ kips})(3 \text{ ft})(12 \text{ in/ft})}{(10 \text{ in}^2)(29{,}000 \text{ kips/in}^2)} = 0.2472 \text{ in}$

$\delta_{Total} \approx 0.25 \text{ in}$

Reference: NCEES *PE Civil Reference Handbook* > Mechanics of Materials > Uniaxial Stress-Strain
Answer: A

SOLUTIONS

17. The free-body diagram shown below is used to calculate F_1 and F_2:

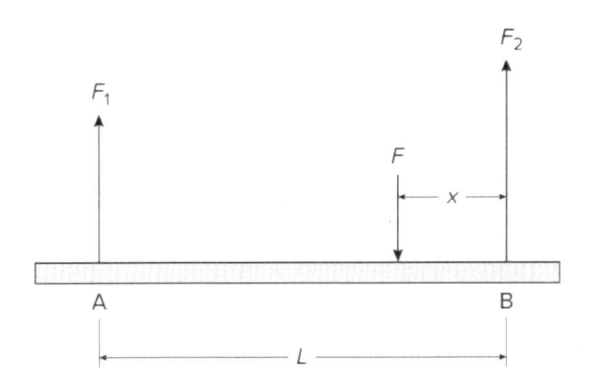

$\sum M_B = 0$

$F_1(4 \text{ m}) - (800 \text{ kN})(1 \text{ m}) = 0$

$F_1 = 200 \text{ kN}$

$\sum F_y = 0$

$F = F_1 + F_2$

$800 \text{ kN} = 200 \text{ kN} + F_2$

$F_2 = 800 \text{ kN} - 200 \text{ kN} = 600 \text{ kN}$

To keep the bar balanced, the deflection of cable 1 is equal to the deflection of cable 2, which is expressed as:

$\delta_1 = \delta_2$

$\delta_1 = \dfrac{F_1 L_1}{E_1 A_1}$

$\delta_2 = \dfrac{F_2 L_2}{E_2 A_2}$

$\dfrac{F_1 L_1}{E_1 A_1} = \dfrac{F_2 L_2}{E_2 A_2}$

Solve for the ratio L_1/L_2 with $E_1 = E_2$:

$\dfrac{L_1}{L_2} = \dfrac{F_2 E_1 A_1}{F_1 E_2 A_2} = \dfrac{(600 \text{ kN})(2{,}600 \text{ mm}^2)}{(200 \text{ kN})(1{,}300 \text{ mm}^2)} = 6$

Reference: NCEES *PE Civil Reference Handbook* > Mechanics of Materials > Uniaxial Stress-Strain

Answer: D

SOLUTIONS

18. The maximum shear force is at the fixed support and equal to the reaction at support A. The reaction at support A is calculated as:

$R_A = wL$
$R_A = (1 \text{ kN/m})(4 \text{ m}) = 4 \text{ kN}$
$R_A = V_{Max} = 4 \text{ kN} = 4,000 \text{ N}$

The following is necessary to calculate the shear stress:

A^* = area above the plane where the shear acts

y = distance from the neutral axis to the centroid of A^*

$Q = A^*y$

The maximum Q is at $h/2$ (mid depth of cross section):

$A^* = \left(\dfrac{bh}{2}\right) = \dfrac{(10 \text{ cm})(25 \text{ cm})}{2} = 125 \text{ cm}^2 = 0.0125 \text{ m}^2$

$y = \dfrac{h}{4} = \dfrac{25 \text{ cm}}{4} = 6.25 \text{ cm} = 0.0625 \text{ m}$

$Q_{Max} = (0.0125 \text{ m}^2)(0.0625 \text{ m}) = 0.0007812 \text{ m}^3$

The moment of inertia (I) is calculated as:

$\dfrac{bh^3}{12} = \left(\dfrac{1}{12}\right)(10 \text{ cm})(25 \text{ cm})^3 = 13,020.8 \text{ cm}^4 = 0.0001302 \text{ m}^4$

Maximum shearing stress is calculated as:

$\tau_{Max} = \dfrac{V_{Max} Q_{Max}}{Ib} = \dfrac{(4,000 \text{ N})(0.0007812 \text{ m}^3)}{(0.0001302 \text{ m}^4)(0.1 \text{ m})}$

$\tau_{Max} = 240,000 \text{ N/m}^2 = 240 \text{ kPa}$

Reference: NCEES *PE Civil Reference Handbook* > Mechanics of Materials > Beams
Answer: A

SOLUTIONS

19. Pipes 1 and 2 are the 18-in pipes, and pipe 3 is the 30-in pipe. The continuity equation states:

$V_1 A_1 + V_2 A_2 = V_3 A_3$

$A_1 = A_2 = \dfrac{\pi D^2}{4} = \dfrac{3.14 \left(\dfrac{18}{12}\right)^2}{4} = 1.77 \text{ ft}^2$

$A_3 = \dfrac{3.14 \left(\dfrac{30}{12}\right)^2}{4} = 4.91 \text{ ft}^2$

$V_1 = 2.5 \text{ ft/s}$

$V_2 = 3.8 \text{ ft/s}$

Solve the continuity equation for V_3:

$V_3 = \dfrac{(V_1 A_1 + V_2 A_2)}{A_3}$

$V_3 = 2.27 \text{ ft/s}$

Reference: NCEES *PE Civil Reference Handbook* > Hydraulics > Principles of One-Dimensional Fluid Flow

Answer: B

SOLUTIONS

20. Manning's equation: $\dfrac{Q = KAR_H^{\frac{2}{3}}S^{\frac{1}{2}}}{n}$

Where:
K = 1.486 for USCS units

For circular pipes flowing full, R_H is $D/4$:

$$Q = \dfrac{1.486\left(\pi\dfrac{D^2}{4}\right)\left(\dfrac{D^{\frac{2}{3}}}{4}\right)\left(S^{\frac{1}{2}}\right)}{n} \rightarrow 2.16\dfrac{nQ}{\sqrt{S}} = D^{\frac{8}{3}}$$

Solve for D:

$$D = 1.335\left(\dfrac{nQ}{\sqrt{S}}\right)^{3/8}$$

$Q = 200$ ft³/s

$n = 0.024$

$S = 0.02$ ft/ft

$$D = 1.335\left(\dfrac{0.024(200)}{(0.02)^{1/2}}\right)^{3/8} = 5 \text{ ft}$$

Reference: NCEES *PE Civil Reference Handbook* > Open Channel Flow > Manning's Equation

Answer: C

SOLUTIONS

21. The Hazen-Williams equation for the average velocity is:

$$V = k_1 C R_H^{0.63} S^{0.54}$$

In the above equation, $k_1 = 1.318$ for USCS units, R_H = hydraulic radius, C = Hazen Williams coefficient, and S = friction slope (head loss per unit pipe length).

$k_1 = 1.318$

$R_H = \dfrac{\frac{24}{12}}{4} = 0.5 \text{ ft}$

$C = 110$

$S = \dfrac{2.4}{400} = 0.006 \text{ ft/ft}$

$V = 1.318(110)(0.5)^{0.63}(0.006)^{0.54}$

$V = 1.318(110)(0.646)(0.063)$

$V = 5.91 \text{ ft/s}$

Reference: NCEES *PE Civil Reference Handbook* > Hydraulics> Closed Conduit Flow and Pumps > Hazen-Williams Equation

Answer: B

SOLUTIONS

22. The depth of flow at the downstream end of a channel where the water descends in a waterfall is the critical depth. Under these conditions, the following expression applies:

$$\frac{Q^2}{g} = \frac{A^3}{T}$$

Where:
A = cross-sectional area
T = top width at the downstream end of the channel (or pipe in this case)

For a full pipe:

$$A_F = \pi \left(\frac{\frac{15}{12}}{2}\right)^2 = 1.23 \text{ ft}^2$$

If the pipe is half full, then the area is half the full value, or $A = 0.61 \text{ ft}^2$.
For a half-full circular pipe, the top width T is equal to the pipe diameter D of 1.25 ft.

Solve the above equation for Q:

$$Q^2 = \frac{gA^3}{T}$$

$$Q^2 = \frac{(32.2 \text{ ft/s}^2)(0.61 \text{ ft}^2)^3}{1.25 \text{ ft}} = 5.85$$

Q = 2.41 ft³/s

Reference: NCEES *PE Civil Reference Handbook* > Hydraulics > Open Channel Flow > Specific Energy
Answer: C

SOLUTIONS

23. Determine the friction losses in the pipe:

Friction losses in pipe $= \dfrac{0.65 \text{ ft}}{100 \text{ ft}} (55 \text{ ft}) = 0.358 \text{ ft}$

Determine the velocity:

$Q = 250 \text{ gpm}(0.134 \text{ ft}^3/\text{gal})\left(\dfrac{1 \text{ min}}{60 \text{ s}}\right) = 0.56 \text{ ft}^3/\text{s}$

Area of the pipe $= \pi \left(\dfrac{4 \text{ in}}{12 \text{ in/ft}}\right)^2 = 0.35 \text{ ft}^2$

$V = \dfrac{Q}{A}$

$V = \dfrac{0.56 \text{ ft}^3/\text{s}}{0.35 \text{ ft}^2} = 1.6 \text{ ft/s}$

Determine the velocity head (h_V):

$h_V = \dfrac{V^2}{2g} = \dfrac{(1.6 \text{ ft/s})^2}{2(32.2 \text{ ft/s}^2)} = 0.04 \text{ ft}$

Determine the friction loss from the velocity head:

$h_M = h_V K$

K values are given in the question statement.

Friction loss in three check valves = (0.04)(0.3)(3) = 0.036 ft

Friction loss in two gate valves = (0.04)(0.15)(2) = 0.012 ft

Sum friction loss:
Total losses = 0.358 ft + 0.036 ft + 0.012 ft = 0.41 ft

Reference: NCEES *PE Civil Reference Handbook* > Hydraulics > Principles of One-Dimensional Fluid Flow

Answer: C

SOLUTIONS

24. Apply the rational formula:

$Q = CiA$

Q = discharge (ft³/s)

C = runoff coefficient for the watershed

i = rainfall intensity (in/hr)

A = watershed area (acres)

When applying the rational method, rain falling over a time period equal to the time of concentration of the watershed should be used. In this case, the time of concentration is given as 40 min, or 0.67 hr, and the corresponding rainfall amount is 0.7 in.

Rainfall intensity:
$$i = \frac{0.7 \text{ in}}{(40 \text{ min})\left(\frac{1 \text{ hr}}{60 \text{ min}}\right)} = 1.04 \text{ in/hr}$$

Solve for the discharge:
$Q = CiA$
$Q = (0.2)(1.04)(3) = 0.63 \text{ ft}^3/\text{s}$

Reference: NCEES *PE Civil Reference Handbook* > Hydrology > Runoff Analysis
Answer: B

25. The length of the curve from PC to PT is:
$$L = \frac{RI\pi}{180°} = \frac{(2{,}480 \text{ ft})(70°)\pi}{180°} \cong 3{,}029.89 \text{ ft}$$

The station of PT is:
sta PT = sta PC+L = (sta 8+60) + 3,029.89 ft
sta PT = (sta 8+60) + (sta 30+29.89) = sta 38+89.89

Reference: NCEES *PE Civil Reference Handbook* > Transportation > Horizontal Design > Basic Curve Elements
Answer: D

SOLUTIONS

26. The jam density occurs when the speed is 0 mph, or:
$v = 55 - 0.4k = 0$

$k = \dfrac{55}{0.45} = 122.2$ veh/mi/ln

The free-flow speed occurs when the density is 0 veh/mi/ln, or:
$v = 55 - 0.45(0) = 55$ mph

Reference: NCEES *PE Civil Reference Handbook* > Transportation > Traffic Engineering > Uninterrupted Flow
Answer: D

27. In horizontal curve formulas, tangent distance is the distance from the PC to PI or from the PI to PT (Answer A is correct).

Answer B is incorrect. The distance from the PI to the middle point of the curve is external distance.

Answer C is incorrect. The distance along the line joining the PC and the PT is the long chord.

Answer D is incorrect. The distance from the middle point of the curve to the middle of the chord joining the PC and PT is the middle ordinate.

Reference: NCEES *PE Civil Reference Handbook* > Transportation > Horizontal Design > Basic Curve Elements
Answer: A

SOLUTIONS

28. Use the arc definition:

$$R = \frac{5{,}729.58}{D}$$

$$R = \frac{5{,}729.58}{7.74°} = 740.3 \text{ ft}$$

Find the tangent length formula:

$$T = R \tan\left(\frac{\Delta}{2}\right)$$

$$\Delta = 13°21'55''$$

Convert to decimal:

$$13° + \left(\frac{21}{60}\right) + \left(\frac{55}{3{,}600}\right) = 13.35°$$

$$T = (740.3 \text{ ft})\left[\tan\left(\frac{13.35°}{2}\right)\right] = 86.6 \text{ ft} \approx 87 \text{ ft}$$

Reference: NCEES *PE Civil Reference Handbook* > Transportation > Horizontal Design > Basic Curve Elements
Answer: A

29. Increasing the water content in concrete will increase the workability (slump), making it easier for placement as expected since this will increase the cement paste content. However, because not all water is consumed during the cement hydration, the final concrete product will have a higher porosity as a result of the additional water. Therefore, the increased water content will compromise the strength in the process.

Reference: NCEES *PE Civil Reference Handbook* > Material Quality Control and Production > Material Properties and Testing
Answer: A, B

SOLUTIONS

30. The relative compaction cannot exceed 100%, so answers C and D can be eliminated. To determine the relative compaction, the dry unit weight of the fill sample must be determined first. Then, compare it to the maximum dry unit weight.

$$\gamma_d = \frac{\frac{W_T}{V_T}}{1 + \frac{w\%}{100\%}}$$

$$\gamma_d = \frac{\frac{62 \text{ lb}}{864 \text{ in}^3 \left(\frac{1 \text{ ft}}{12 \text{ in}}\right)^3}}{1 + \frac{15\%}{100\%}} = \frac{\frac{62 \text{ lb}}{0.5 \text{ ft}^3}}{1.15}$$

$$\gamma_d = 107.8 \text{ lb/ft}^3$$

The relative compaction is the dry unit weight of the sample divided by the maximum unit weight:

$$100\% \left(\frac{\gamma_d}{\gamma_{d,\max}}\right) = 100\% \left(\frac{107.8 \text{ lb/ft}^3}{115 \text{ lb/ft}^3}\right) = 93.8\%$$

Reference: NCEES *PE Civil Reference Handbook* > Geotechnical > Weight-Volume Relationships
Answer: A

31. The soil below the water table is saturated, and $\gamma = \gamma_{Sat} = \gamma_{Dry}(1 + w)$.

$\gamma_{Sat} = 95 \text{ lb/ft}^3 (1 + 0.25) = 118.8 \text{ lb/ft}^3$

$\gamma_b = \gamma_{Sat} - \gamma_w$

$\gamma_b = 118.8 \text{ lb/ft}^3 - 62.4 \text{ lb/ft}^3 = 56.4 \text{ lb/ft}^3$

Reference: NCEES *PE Civil Reference Handbook* > Geotechnical > Weight-Volume Relationships
Answer: B

SOLUTIONS

32. Assume that the volume of soil solids (V_S) is 1 ft³. The void ratio (e) is 0.7. The volume of water and air ($V_V = eV_S$) is 0.7 ft³.

The weight of the soil solids is:
$W_S = V_S G \gamma_W = (1 \text{ ft}^3)(2.7)(62.4 \text{ lb/ft}^3) = 168.48 \text{ lbf}$

The weight of water is:
$W_W = wW_S = 22\%(168.48 \text{ lb}) = 37.0656 \text{ lb}$

The weight of water and soil solids is:
$W = W_S + W_W = 168.48 \text{ lb} + 37.0656 \text{ lb} = 205.55 \text{ lb}$

Plot the known information in the phase diagram.

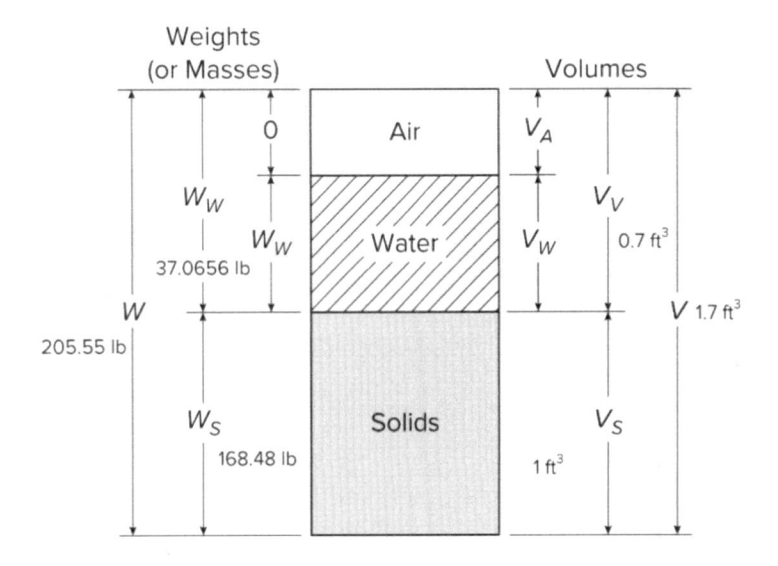

The total unit weight of the soil is:
$$\gamma = \frac{W}{V} = \frac{205.55 \text{ lb}}{1.7 \text{ ft}^3} \cong 120.9 \text{ lb/ft}^3$$

Reference: NCEES *PE Civil Reference Handbook* > Geotechnical > Weight-Volume Relationships
Answer: C

33. Type V cement is sulfate-resisting and is specified when there is extensive exposure to sulfates. This includes water with high alkali content and seawater.

Answer: D

SOLUTIONS

34. Use the "Properties of NZ and PZ Steel Sheet Piles" table in chapter 4, section 4.2.2 of the NCEES *PE Civil Reference Handbook*:

Plastic section modulus of PZ 22 = 21.79 in³/ft.

Plastic section modulus of PZ 35 = 57.17 in³/ft.

Therefore, the ratio is:

$$\frac{21.79}{57.17} = 0.38$$

Reference: NCEES *PE Civil Reference Handbook* > Steel Design > PZ Hot-Rolled Steel Sheet Pile Properties
Answer: B

35. According to the "Nondestructive Test Methods for Concrete" table in chapter 2, section 2.5.3.2 of the NCEES *PE Civil Reference Handbook*, internal defects can be identified by impact-echo, ultrasonic echo, and impulse response. Therefore, A and E are correct.

Reference: NCEES *PE Civil Reference Handbook* > Material Quality Control and Production > Concrete Maturity and Early Strength Evaluations
Answer: A, E

36. Recognize that the constructed condition is in compacted measure. Calculate the volume (compacted measure) of the roadway: $V = AL$

The cross section is a trapezoid with a top width of 50 ft. The height is 4.5 ft, and the slope is 3:1 (H:V).

Therefore, the bottom width is 50 ft + 2[3(4.5 ft)] = 77 ft.

$$A = \left[\left(\frac{50 \text{ ft} + 77 \text{ ft}}{2}\right)(4.5 \text{ ft})\right] = 285.75 \text{ ft}^2$$

Calculate volume:

$$V = (285.75 \text{ ft}^2)(5 \text{ miles})\left(\frac{5{,}280 \text{ ft}}{1 \text{ mile}}\right)\left(\frac{1 \text{ yd}^3}{27 \text{ ft}^3}\right) = 279{,}400 \text{ yd}^3$$

Convert compacted volume into bank volume:

$$V_{\text{Compacted}} = (1 - S_H)V_{\text{Bank}}$$
$$279{,}400 \text{ yd}^3 = 0.80 V_{\text{Bank}}$$
$$V_{\text{Bank}} = 349{,}250 \text{ yd}^3$$

Reference: NCEES *PE Civil Reference Handbook* > Construction > Earthwork Construction and Layout > Excavation and Embankment
Answer: C

SOLUTIONS

37. Between sta 1+25 and sta 3+50, the length is 225 ft. The volume of cut and fill are calculated as:

$$V = \left(\frac{A_1 + A_2}{2}\right)L$$

$$V_{Cut} = \left(\frac{20\ ft^2 + 230\ ft^2}{2}\right)\left(\frac{225\ ft}{27}\right) = 1{,}041.67\ yd^3$$

$$V_{Fill} = \left(\frac{110\ ft^2 + 50\ ft^2}{2}\right)\left(\frac{225\ ft}{27}\right) = 666.67\ yd^3$$

$$V_{Net} = V_{Cut} - V_{Fill} = 1{,}041.67\ yd^3 - 666.67\ yd^3 = 375\ yd^3\ (cut)$$

Reference: NCEES *PE Civil Reference Handbook* > Construction > Earthwork Construction and Layout > Earthwork Volumes
Answer: A

38. N = number of injuries, illnesses, and fatalities = 6 + 3 + 1 = 10

T = total hours worked by all employees during the period in question
T = 135[56(50)] = 378,000 hours

$$IR = \frac{200{,}000N}{T} = \frac{10(200{,}000)}{378{,}000} = 5.29$$

Reference: NCEES *PE Civil Reference Handbook* > Health and Safety > Safety Management and Statistics
Answer: C

39. Type C over type B soil requires slopes (H:V) of (1.5:1) and (1:1), respectively.

Type C = 1.5(6 ft) = 9 ft
Type B = 1(7 ft) = 7 ft

Total horizontal distance = undisturbed perimeter + horizontal distance associated with type C layer + horizontal distance associated with type B layer

Total horizontal distance = 3 ft + 9 ft + 7 ft = 19 ft

Reference: NCEES *PE Civil Reference Handbook* > Trench and Construction Safety > Slope Configurations: Excavations in Layered Soils
Answer: B

SOLUTIONS

40. To compute the difference in elevation from (2) points, compute the difference in the sum of both the Backsights and Foresights between the same (2) points, respectively.

Elevation of BM_1 + BS = HI → 12.31 ft + 5.43 ft = 17.74 ft

Elevation of TP_1 = HI − FS = 17.74 ft − 2.31 ft = 15.43 ft

Elevation of TP_1 + BS = HI → 15.43 ft + 4.72 ft = 20.15 ft

Elevation of TP_2 = HI − FS = 20.15 ft − 1.42 ft = 18.73 ft

Elevation of TP_2 + BS = HI → 18.73 ft + 6.09 ft = 24.82 ft

Elevation of BM_2 = HI − FS = 24.82 ft − 1.86 ft = 22.96 ft

Difference in elevation of BM_2 and BM_1 = 22.96 ft − 12.31 ft = 10.65 ft

The calculations are summarized in the table below:

POINT	BS (+)	HI	FS (−)	ELEVATION
BM_1	5.43	17.74		12.31
TP_1	4.72	20.15	2.31	15.43
TP_2	6.09	24.82	1.42	18.73
BM_2			1.86	22.96
	+16.24		−5.59	+10.65

Reference: NCEES *PE Civil Reference Handbook* > Construction > Earthwork Construction and Layout > Site Layout and Control

Answer: B

SOLUTIONS

41. Roundabouts have less potential conflict points than conventional intersections. In the case of the four-legged intersection, 32 conflict points are replaced by 8, as illustrated in the figure below:

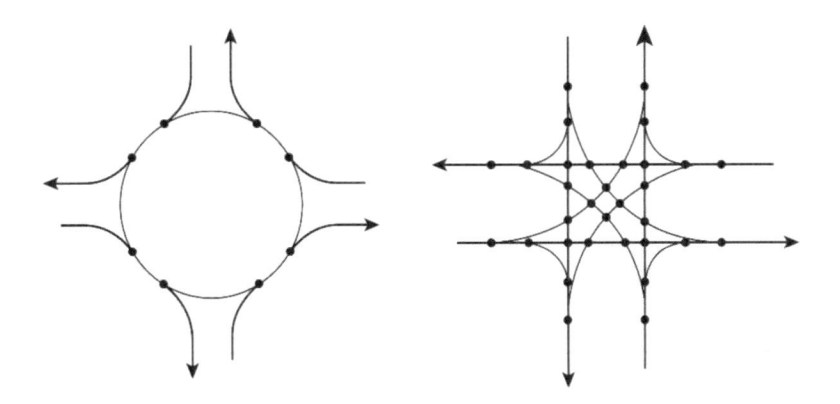

Reference: AASHTO *Policy on Geometric Design of Highways and Streets* (*Green Book*), 7th edition > Chapter 9 > Figure 9.1
Answer: A

42. The utilities of driving alone, carpooling, and bus are:
$$U_A = -(0.50) - \frac{5(3.00)}{50} = -0.80$$
$$U_C = -(0.75) - \frac{5(1.20)}{50} = -0.87$$
$$U_B = -(1.00) - \frac{5(0.80)}{50} = -1.08$$
The probability that an individual would select carpooling to work is:
$$P_C = \frac{e^{U_C}}{e^{U_A} + e^{U_C} + e^{U_B}} = \frac{e^{-0.87}}{e^{-0.80} + e^{-0.87} + e^{-1.08}} \cong 0.347$$

Reference: *Highway Capacity Manual* > Volume 1 > Chapter 6
Answer: C

43. The signalized intersection can be controlled by signals operating in two or more phases. A two-phase signal has one phase for each axis of travel. A three-phase signal provides one of the roads with a left-turn phase. A four-phase signal allows both roads with left-turn phases.

References: *Highway Capacity Manual* > Volume 3 > Chapter 19
Answer: B, C, D

SOLUTIONS

44. Use the equation for space mean speed.

$$\text{SMS} = \frac{nL}{\sum_1^n t_i} = \frac{4(300)}{10 + 7.5 + 6 + 5} = 42.1 \text{ ft/s}$$

Reference: NCEES *PE Civil Reference Handbook* > Traffic Engineering > Space Mean Speed
Answer: B

45. Let U_B, U_C, and U_T represent the utilities of bus, car, and train, respectively. Based on the given logit model, the probability that Scott will select the train to go to work is:

$$P_T = \frac{e^{U_T}}{e^{U_B} + e^{U_C} + e^{U_T}}$$

For this problem, $U_B = -0.02, U_C = -0.3,$ and $U_T = -0.01$.
Therefore:

$$P_T = \frac{e^{U_T}}{e^{U_B} + e^{U_C} + e^{U_T}} = \frac{e^{-0.01}}{e^{-0.02} + e^{-0.3} + e^{-0.01}} \cong 0.365$$

Reference: *Highway Capacity Manual* > Volume 1 > Chapter 6
Answer: B

46. The LOS for pedestrians is used to evaluate the performance of a sidewalk, change the width of a sidewalk, and remove the street furniture.

Reference: *Highway Capacity Manual* > Volume 1 > Chapter 5
Answer: A, C, D

47. Due to the high speed and volume of traffic on freeways, grade separations without ramps or interchanges should be employed to safely and efficiently accommodate traffic safely and efficiently through intersections. Interchange is the appropriate choice because there is turning traffic between the two freeways.

Reference: AASHTO *Policy on Geometric Design of Highways and Streets* (*Green Book*), 7th edition > Chapter 10
Answer: C

SOLUTIONS

48. The left-turn crash rate per million entering vehicles (RMEV) is calculated as:

$$RMEV = \frac{A(1{,}000{,}000)}{ADT(365)}$$

Where:

A = number of left-turn crashes occurring in a single year at the intersection
ADT = average daily traffic entering the intersection
For this problem, $A = 6$ and ADT $= 1{,}000$.

Therefore:

$$RMEV = \frac{6(1{,}000{,}000)}{1{,}000(365)} \cong 16.4$$

Reference: *Highway Safety Manual* > Volume 1 > Chapter 4
Answer: B

49. The formula to determine the vehicle signal change interval (red interval) is:

$$r = \frac{W + l}{v}$$

Where:
r = length of red clearance interval to the nearest 0.1 second
v = vehicle approach speed (ft/s)
W = curb-to-curb width of intersection (ft)
l = length of vehicle (ft)

Adjust the equation and solve for W:
$W = rv - l$
$W = (1.8 \text{ s})(37.8 \text{ ft/s}) - 20 \text{ ft} \cong 48 \text{ ft}$

Reference: *Highway Capacity Manual* > Volume 3 > Exhibit 19-7
Answer: D

SOLUTIONS

50. The following equation is used to calculate the proportion of vehicles arriving during green:

$$P = R_P \left(\frac{G}{C}\right)$$

$$P = 1\left(\frac{29 \text{ s}}{60 \text{ s}}\right) = 0.48$$

Reference: *Highway Capacity Manual* > Volume 3 > Chapter 19 > Equation 19-5
Answer: C

51. Use the given average daily traffic to calculate the annual traffic:
V = ADT(365)

Where:
V = annual traffic entering intersection
ADT = average daily traffic entering intersection

Therefore:
V = 300(365) = 109,500

Reference: *Highway Capacity Manual* > Volume 3 > Chapter 19
Answer: B

52. First, establish chord $\overline{A - B}$.

Find the radius:
$$R = \frac{5{,}729.578}{D} = \frac{5{,}729.578}{6.75°} = 848.826$$

Find arc length BX = sta 20 + 00.00 − sta 16 + 50.00 = 350 ft

Establish chord $\overline{BX}$.

Find angle β and use it to find angle α. Note that $\alpha = \frac{\beta}{2}$.

Use angle α to find the bearing of chord $\overline{BX}$.

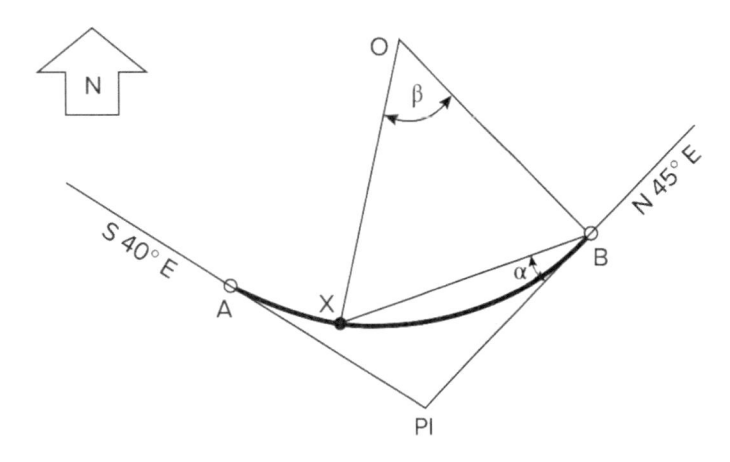

Find the bearing and length of chord $\overline{BX}$.

Notice that triangle X-O-B has two sides of length R separated by angle β.

$$\text{Angle } \beta = \left(\frac{350'}{2\pi R}\right)(360°) = \left(\frac{350'}{2\pi(848.826')}\right)(360°) = 23.625°$$

$$\text{Angle } \alpha = \frac{23.625°}{2} = 11.813°$$

Azimuth of chord $\overline{BX}$ azimuth of ahead tangent bearing + 180° + angle α
= Ax 45° + 180° + 11.813° = Az 236.813°

Bearing of chord $\overline{BX}$ = Az 236.813° − 180° = S 56.813°W

$$\text{Length of chord } \overline{BX} = (2)(848.826')\left(\sin\frac{23.625°}{2}\right)$$
$$= (2)(848.826')(\sin 11.813°) = 347.526'$$

Use latitudes and departures of chord $\overline{BX}$ to determine coordinates of point X. Chord $\overline{BX}$ has a SW bearing. Therefore, both the latitudes and departure are negative.

Latitude is change in north-south direction:
= Length · cos (bearing angle) = (347.526')(cos 56.813°) = −190.23'

SOLUTIONS

Departures is change in east – west direction:
= Length · sin (bearing angle) = (347.526′)(sin 56.813°) = −290.84′

Coordinates of point X
Point B N coordinate + Latitude: 10,000.00 − 190.23′ = N 9,809.77
Point B E coordinate + Departure: 34,000.00 − 290.84′ = E 33,709.16

References: NCEES *PE Civil Reference Handbook* > Horizontal Design > Basic Curve Elements
Answer: C

53. The problem requires the calculation of the tangent length (T) and length of the curve (L) to then obtain the station location of point of curvature (PC) and the point of tangent (PT) location.

Obtain radius (R) by converting degrees to decimal:

$$D = 9°30' = 9 + \frac{30}{60} = 9.5$$

$$R = \frac{5{,}729.578}{D} = \frac{5{,}729.578}{9.5} = 603 \text{ (ft)}$$

Obtain length of tangent (T) and length of curve (L):

$$T = R \tan\left(\frac{1}{2}\right) = 603\left[\tan\left(\frac{40.5}{2}\right)\right] = 222.46 \text{ (ft)} = \text{sta } 2 + 22.46$$

$$L = RI\left(\frac{\pi}{180}\right) = 603\left[40.5\left(\frac{\pi}{180}\right)\right] = 426.24 \text{ (ft)}$$

Calculate PC and PT (convert T and L to station):
PC = PI − T = (21 + 00) − (2 + 22.46) = 18 + 77.54
PT = PC + L = (18 + 77.54) + (4 + 26.24) = 23 + 03.78

Reference: NCEES *PE Civil Reference Handbook* > Horizontal Design > Basic Curve Elements
Answer: C

SOLUTIONS

54. Use the horizontal sight line offset equation:

$$\text{HSO} = R\left[1 - \cos\left(\frac{28.65 S}{R}\right)\right]$$

$$\text{HSO} = 45 \text{ ft}\left[1 - \cos\left(\frac{28.65(40 \text{ ft})}{45 \text{ ft}}\right)\right] = 4.4 \text{ ft}$$

Reference: AASHTO *Policy on Geometric Design of Highways and Streets* (*Green Book*), 7th edition > Equation 3-37
Answer: D

55. According to the problem statement, 60 km/hr is 60% of the design speed. Find the design speed.

$$V = \frac{60 \text{ km/hr}}{(0.6)} = 100 \text{ km/hr}$$

According to *Green Book* Equation 3-8:

$$R = \frac{V^2}{127(0.01e + f)}$$

$$R = \frac{100^2}{127[0.01(8) + 0.15]} = 342 \text{ m}$$

Reference: AASHTO *Policy on Geometric Design of Highways and Streets* (*Green Book*), 7th edition > Equation 3-8
Answer: B

56. The question indicates standard conditions, so the following equation for vertical curves is used:

$$L = \frac{AS^2}{2{,}158}$$

$$L = \frac{6(500 \text{ ft})^2}{2{,}158} = 695.1 \text{ ft}$$

Reference: AASHTO *Policy on Geometric Design of Highways and Streets* (*Green Book*), 7th edition > Equation 3-44
Answer: C

SOLUTIONS

57. Refer to *Green Book* Equation 3-3.

SSD is calculated as:
$$SSD = 1.47Vt + \frac{V^2}{30\left[\left(\frac{a}{32.2}\right) \pm G\right]}$$

Where:
V = design speed (mph)
t = driver reaction time (s)
a = deceleration rate (ft/s²)
G = percent grade divided by 100

Based on the given information:
$$1.47Vt = SSD - \frac{V^2}{30\left[\left(\frac{a}{32.2}\right) \pm G\right]}$$
$$= 855 \text{ ft} - \frac{(50 \text{ mph})^2}{30\left[\left(\frac{4 \text{ ft/s}^2}{32.2}\right) + 0\right]} = 184.2$$

For a 3% upgrade, the stopping sight distance SSD is calculated as:
$$SSD = 184.2 + \frac{(50 \text{ mph})^2}{30\left[\left(\frac{4 \text{ ft/s}^2}{32.2}\right) + 3\%\right]} = 724.5 \text{ ft}$$

When the percent grade increases from 0 to 3%, the SSD will decrease, or become less than 855 ft. Among the four answers, only option A meets this requirement.

Reference: AASHTO *Policy on Geometric Design of Highways and Streets* (*Green Book*), 7th edition > Equation 3-3
Answer: A

58. Refer to *Green Book* Equation 3-53.
$$L = 2S - \frac{800}{A}\left(C - \frac{h_1 + h_2}{2}\right)$$
$$L = 2(1{,}500 \text{ ft}) - \frac{800}{3}\left(15 \text{ ft} - \frac{3 \text{ ft} + 5 \text{ ft}}{2}\right) = 66.7 \text{ ft}$$

Reference: AASHTO *Policy on Geometric Design of Highways and Streets* (*Green Book*), 7th edition > Equation 3-53
Answer: D

SOLUTIONS

59. Refer to the stopping sight distance formula in *Green Book* Equation 3-3.

$$S = \frac{1.47Vt + (V^2)}{30\left[\left(\frac{a}{32.2}\right) \pm G\right]}$$

G in the above formula is either added or subtracted depending on the situation.

Enter the values given and solve for S.

The denominator of the second part of the equation is calculated as:

$30[(f) + G]$

$30[(0.27) - 0.032] = 7.14$

$S = 1.47[(75 \text{ mph})(1.5 \text{ s})] + \left(\frac{75^2}{7.14}\right)$

$S = 953.2 \text{ ft}$

The driver would not hit the obstruction because the sign is 1,000 ft away from the obstruction and the vehicle comes to a stop after 953 ft.

Reference: AASHTO *Policy on Geometric Design of Highways and Streets (Green Book)*, 7th edition > Equation 3-3
Answer: A

60. The *Green Book* states: "When the distance between the successive noses is less than 450 m [1,500 ft], the speed-change lanes should be connected to provide an auxiliary lane."

Reference: AASHTO *Policy on Geometric Design of Highways and Streets (Green Book)*, 7th edition > Chapter 10 > Section 10.9.6.4.6 Distance between Successive Ramp Terminals
Answer: A

61. The *Green Book* states: "Refuge islands should be a minimum of 6 ft [1.8 m] wide and pedestrians and bicyclists should have a clear path through the island and should not be obstructed by curbs, poles, sign posts, utility boxes, etc."

Reference: AASHTO *Policy on Geometric Design of Highways and Streets (Green Book)*, 7th edition > Chapter 9 > 9.6.3.4 Refuge Islands
Answer: B

SOLUTIONS

62. The problem states that the driver is attempting to make a left turn from stop. Use the *Green Book* intersection control Case B1 (left turn from the minor road) to solve the problem.

Use the notes given in *Green Book* Table 9-6 to adjust the time gap for making left turns on multilane highways.

The problem states that the major road is a four-lane highway. The notes beneath Table 9-6 state that 0.5 s should be added for passenger cars to the time gap (t_g) for each additional lane to be crossed in excess of one. Two additional lanes will be crossed.

The adjustment for each additional line to be crossed is 0.5 s

Use the notes given in Table 9-6 to adjust the time gap for the approach grade of the minor road.

The problem states that the minor road approach grade is 4%. The notes beneath Table 9-6 state that 0.2 s should be added for each percent grade for left turns if the approach grade exceeds 3%.
Adjustment for minor road approach grade = (4)(0.2) = 0.8 s
Determine the adjusted time gap.
$t_{g(\text{Adjusted})} = 7.5 + 0.5 + 0.8 = 8.8$ s

Determine the intersection sight distance (ISD) using *Green Book* Equation 9-1:
$$\text{ISD} = 1.47 V_{\text{Major}} t_g$$
Where:
ISD = intersection sight distance (length of leg of sight triangle along the major road) (ft)
V_{Major} = design speed of major road (mph)
t_g = time gap for minor road vehicle to enter the major road (s)

ISD = 1.47(50 mph)(8.8 s) = 646.8 ft (≈ 647 ft)

Reference: AASHTO *Policy on Geometric Design of Highways and Streets* (*Green Book*), 7th edition > Chapter 9 > Table 9-6
Answer: D

SOLUTIONS

63. Refer to *Green Book* Table 10-6. The design speed of the freeway is 60 mph and the design speed of the exit curve, *V'* is 40 mph.
V = design speed of highway (mph)
V_a = average running speed on highway (mph)
V' = design speed of exit curve (mph)
V'_a = average running speed on exit curve (mph)

Adjust for the grade using *Green Book* Table 10-5. The problem states that this is a deceleration lane with a 4% downgrade.
The ratio from this table multiplied by the length in Table 10-6 gives the length of speed change lane on grade.
$L_{\text{Adjusted}} = (350)(1.2) = 420 \text{ ft}$

Reference: AASHTO *Policy on Geometric Design of Highways and Streets* (*Green Book*), 7th edition > Chapter 10 > Tables 10-5 and 10-6
Answer: D

64. Rearrange the equation from AASHTO *Roadside Design Guide* table 8-9 to solve for the mass of the container:

$$V_1 = \frac{M_V V_0}{M_V + M_l}$$

$$M_l = \frac{M_V V_0}{V_1} - M_V$$

$$M_l = \frac{(4{,}000)(76)}{70} - 4{,}000 = 342.86 \text{ lb} \rightarrow 343 \text{ lb}$$

Reference: AASHTO *Roadside Design Guide* > Chapter 8 > Table 8-9
Answer: B

65. The question indicates parallel installation, so the reduced equation is used:

$$X = \frac{(L_A) - (L_2)}{L_A / L_R}$$

$$X = \frac{(25) - (15)}{25/300} = 120 \text{ ft}$$

Answer: A
Reference: AASHTO *Roadside Design Guide* > Chapter 5 > Section 5.6.4 Length-of-Need

SOLUTIONS

66. New Jersey, low profile, and constant slope are all types of concrete barriers.

Reference: AASHTO *Roadside Design Guide* > Chapter 5 > Section 5.4.1.12 Concrete Barriers
Answer: B, E, F

67. Maximum curb ramp grade for accessibility should be 8.33%.

Reference: AASHTO *Policy on Geometric Design of Highways and Streets* (*Green Book*), 7th edition > Chapter 4 > Section 4.17.3 Curb Ramp Design
Answer: B

68. Review the criteria in *MUTCD*, section 4C.10 to determine if a signal warrant is needed.

Use *MUTCD* Figure 4C-10 below. This is the same diagram shown in the problem—two approach lanes at the track crossing.

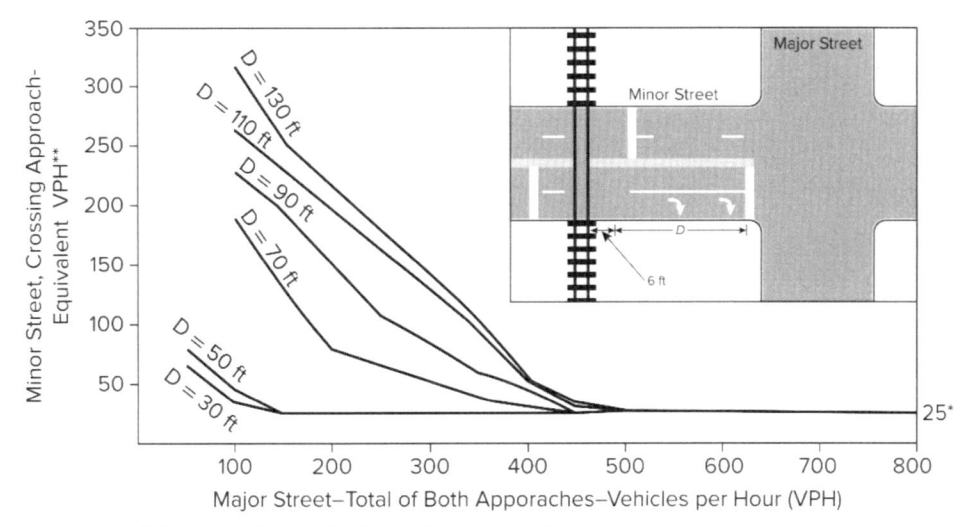

*25 vph applies as the lower threshold volume
**VPH after applying the adjustment factors in tables 4C-2, 4C-3, and/or 4C-4, if appropriate

Source: Federal Highway Administration (FHWA). 2009. *Manual on Uniform Traffic Control Devices* (*MUTCD*). Washington, DC: US Department of Transportation.

The major street (both approaches) traffic volume is 250 vehicles per hour. The minor street (one approach) traffic volume is 150 vehicles per hour.

SOLUTIONS

However, the minor street traffic volumes must be adjusted based on the information given in the problem. Use *MUTCD* Tables 4C-2, 4C-3, and 4C-4 (shown below) to make the adjustments.

TABLE 4C-2. Warrant 9, Adjustment Factor for Daily Frequency of Rail Traffic

RAIL TRAFFIC PAR DAY	ADJUSTMENT FACTOR
1	0.67
2	0.91
3 to 5	1.00
6 to 8	1.18
9 to 11	1.25
12 or more	1.33

TABLE 4C-3. Warrant 9, Adjustment Factor for Percentage of High-Occupancy Buses

% OF HIGH-OCCUPANCY BUSES* ON MINOR-STREET APPROACH	ADJUSTMENT FACTOR
0%	1.00
2%	1.09
4%	1.19
6% or more	1.32

*A high-occupancy bus is defined as a bus occupied by at least 20 people.

TABLE 4C-4. Warrant 9, Adjustment Factor for Percentage of Tractor-Trailer Trucks

% OF TRACTOR-TRAILER TRUCKS ON MINOR-STREET APPROACH	ADJUSTMENT FACTOR D LESS THAN 70 FEET	ADJUSTMENT FACTOR D OF 70 FEET OR MORE
0% to 2.5%	0.50	0.50
2.6% to 7.5%	0.75	0.75
7.6% to 12.5%	1.00	1.00
12.6% to 17.5%	2.30	1.15
17.6% to 22.5%	2.70	1.35
22.6% to 27.5%	3.28	1.64
More than 27.5%	4.18	2.09

Source: MUTCD (Public domain)

Source: Federal Highway Administration (FHWA). 2009. *Manual on Uniform Traffic Control Devices* (*MUTCD*). Washington, DC: US Department of Transportation.

The problem states that there are one rail traffic/day, 2% high-occupancy buses, and 20% tractor trailer trucks. From the charts above, the adjustment to the minor street vehicular traffic is as follows:

Minor street traffic volume$_{Adj}$ = (150 vph)(0.67)(1.09)(1.35) = 147.86 vph

SOLUTIONS

Plot the major street (both approaches) and minor street (one approach) traffic volumes adjusted using *MUTCD* Figure 4C-10. The distance, *D* = 95 ft, from the stop line to the center of the track nearest the intersection is not shown in the figure. According to the *MUTCD* section 42.10, "…the curve for the distance *D* that is nearest to the actual distance *D* should be used." Therefore, the curve showing *D* = 90 ft should be used as shown.

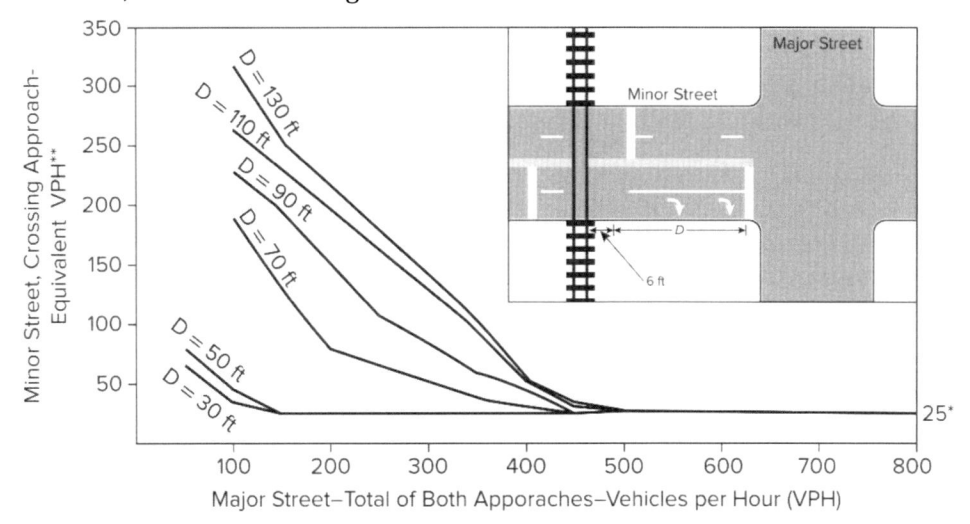

*25 vph applies as the lower threshold volume

**VPH after applying the adjustment factors in tables 4C-2, 4C-3, and/or 4C-4, if appropriate

Source: Federal Highway Administration (FHWA). 2009. *Manual on Uniform Traffic Control Devices* (*MUTCD*). Washington, DC: US Department of Transportation.

According to *MUTCD* Figure 4C-10, the major street (both approaches) vehicular traffic and minor street (one approach) vehicular traffic, and *D* (distance from the stop line to the center of the tract nearest the intersection) plot above the applicable curve. A traffic control signal is warranted.

Reference: *Manual on Uniform Traffic and Control Devices for Streets and Highways* (*MUTCD*) 2009 > Chapter 4C > Section 4C.10
Answer: 148

SOLUTIONS

69. Determine the crash rate for each intersection.

$$R_{Int} = \frac{A \times 10^6}{(T)(V)365}$$

Where:

R_{Int} = crash rate for the intersection
A = number of accidents
T = time period (years)
V = annual average daily traffic volume entering the intersection (vpd)

For intersection 1: $R_{Int}\,1 = \dfrac{5 \times 10^6}{(0.66667)(8,000)(365)} = 2.57$

For intersection 2: $R_{Int}\,2 = \dfrac{25 \times 10^6}{(0.66667)(26,500)(365)} = 3.88$

For intersection 3: $R_{Int}\,3 = \dfrac{34 \times 10^6}{(0.66667)(42,300)(365)} = 3.30$

For intersection 4: $R_{Int}\,4 = \dfrac{16 \times 10^6}{(0.66667)(19,000)(365)} = 3.46$

Intersections 2 and 4 should be improved because they have the highest crash rates.

Reference: *Highway Safety Manual* > Chapter 4
Answer: C

70. The effective green time is:

$g = G + Y - L_2 = 56\text{ s} + 4\text{ s} - 1\text{ s} = 59\text{ s}$

In each signal cycle of 90 s, the vehicles on approach B have g = 59 s to traverse the intersection. Therefore, the capacity of approach B at the signalized intersection (the maximum number of vehicles on approach B that can traverse the intersection in one hour) is:

$C = s\left(\dfrac{g}{C}\right) = (1{,}600 \text{ veh/hr} \cdot \text{lane})\left(\dfrac{59\text{ s}}{90\text{ s}}\right) = 1{,}049 \ \dfrac{\text{veh}}{\text{hr} \cdot \text{lane}}$

Reference: *Highway Capacity Manual* > Volume 3 > Chapter 19
Answer: D

SOLUTIONS

71. *MUTCD* section 6F.68 states: "Stripes on barricade rails shall be alternating orange and white retroreflective stripes sloping downward at an angle of 45 degrees in the direction road users are to pass."

Reference: *Manual on Uniform Traffic and Control Devices for Streets and Highways* (*MUTCD*), 2009 > Chapter 6F > Section 6F.68
Answer: B

72. The problem states that this work would take place in the center of an intersection. The *MUTCD* has 46 typical applications for a variety of situations that could be encountered for temporary work in the public right-of-way. Figure 6H-26 best describes the situation in the problem as shown in the figure.

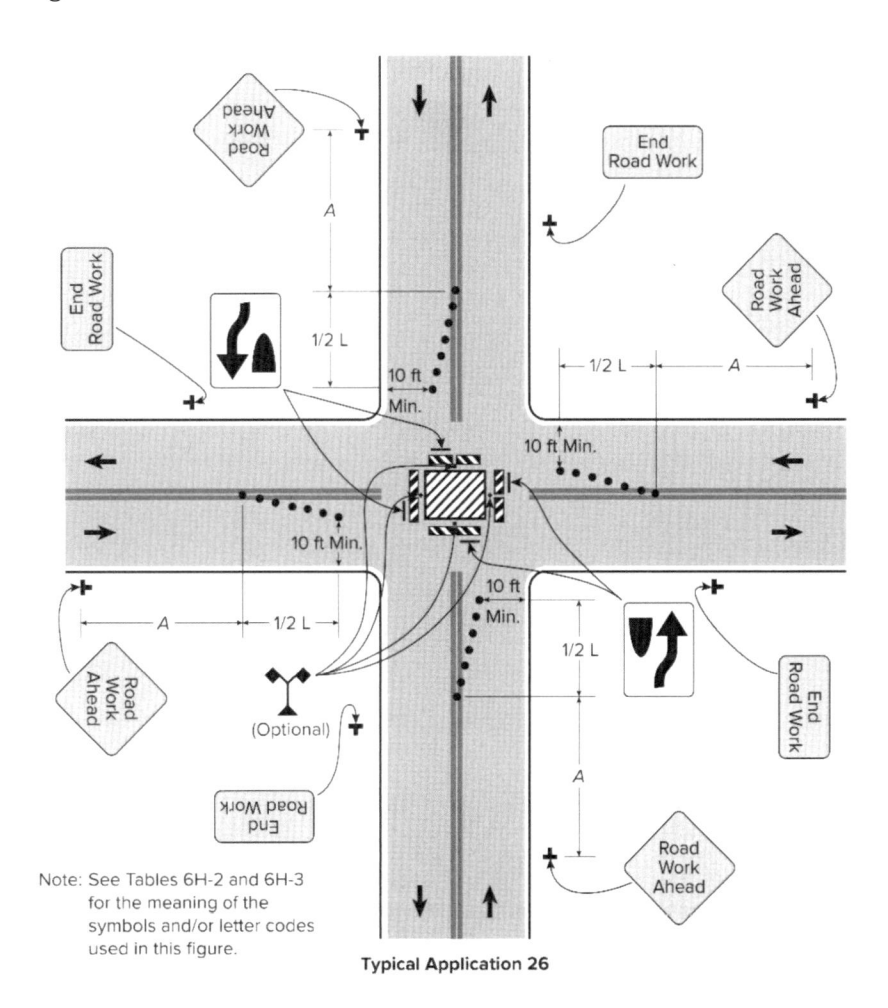

Typical Application 26

Source: Federal Highway Administration (FHWA). 2009. *Manual on Uniform Traffic Control Devices* (*MUTCD*). Washington, DC: US Department of Transportation.

SOLUTIONS

The minimum taper length required to channelize traffic away from the center of the intersection can be found using the chart below.

SPEED (S)	TAPER LENGTH (L) IN FEET
40 mph or less	$L = \dfrac{WS^2}{60}$
45 mph or more	$L = WS$

Source: MUTCD (Public domain)

Where:
L = taper length (ft)
W = width of offset (ft)
S = posted speed limit, or off-peak 85th percentile speed prior to work starting, or the anticipated operating speed

This is an urban low-speed facility with a posted speed limit of 30 mph. Use the following equation to determine L. Generally, lane widths are assumed to be 12 ft unless otherwise stated.

$$L = \frac{WS^2}{60} = \frac{(12 \text{ ft})(30 \text{ mph}^2)}{60} = 180 \text{ ft}$$

The taper shown in *MUTCD* Figure 6H-26 is known as a shifting taper, which is used when traffic is expected to make a lateral shift around obstructions (in this case, around the workspace). As noted in *MUTCD* Table 6C-3 (shown below), shifting tapers are 0.5 L.

TYPE OF TAPER	TAPER LENGTH
Merging Taper	at least L
Shifting Taper	at least 0.5 L
Shoulder Taper	at least 0.33 L
One-Lane, Two-Way Traffic Taper	50 feet minimum, 100 feet maximum
Downstream Taper	50 feet minimum, 100 feet maximum

Source: MUTCD (Public domain)

Therefore, the required length of the taper is: $\left(\dfrac{1}{2}\right)(180) = 90$ ft

References: *Manual on Uniform Traffic and Control Devices for Streets and Highways* (*MUTCD*), 2009 > Chapter 6H
AASHTO *Policy on Geometric Design of Highways and Streets* (*Green Book*), 7th edition > Chapter 4 > Section 4.3
Answer: A

SOLUTIONS

73. According to *MUTCD* section 3B.07:

01: "Edge line markings shall be placed on paved streets or highways with the following characteristics:
    A. Freeways,
    B. Expressways, and
    C. Rural arterials with a traveled way of 20 feet or more in width and an ADT of 6,000 vehicles per day or greater."

02: "Edge lines markings should be placed on paved streets or highways with the following characteristics:
    A. Rural arterials and collectors with a traveled way of 20 feet or more in width and an ADT of 3,000 vehicles per day or greater.
    B. At other paved streets and highways where an engineering study indicates a need for edge line markings."

05: "Edge line markings may be excluded, based on engineering judgement, for reasons such as if the traveled way edges are delineated by curbs, parking, or other markings."

Reference: *Manual on Uniform Traffic and Control Devices for Streets and Highways* (*MUTCD*) 2009 > Chapter 3B > Section 3B.07
Answer: C

74. The required SN is 4.0.

The SN provided by the pavement structure is:
$a_1 D_1 + a_2 D_2 = (0.40)(6) + (0.16) D_2$

Let the provided SN equal the required SN.
$(0.40)(6) + (0.16) D_2 = 4.0$

Solve for the thickness of AB.
$D_2 = 10 \text{ in}$

Reference: AASHTO *Design of Pavement Structures* > Chapter 3 > Section 3.1.4
Answer: D

SOLUTIONS

75. The ESAL is an AASHTO-developed concept to quantify the total amount of loading that a given road is set to experience. Apply the ESAL formula (below) for year 0 (2019) and year 4 (2023) and get the difference in ESALs.

ESAL = ADT$(T)(F_T)(G)(Y)(D)(L)$365

Where:
ADT = 3,500
T = % trucks, which is 1 because in this case the ADT is all trucks
F_T = truck factor
$G(Y)$ = growth factor applied over Y number of years; obtained using F/A engineering economic formula (F/P, $i\%$, n)
D = directional factor
L = lane distribution factor

$G(Y)$ for 2019 = 1 (year zero)
$G(Y)$ for 2023 = (F/A, 5%, 4 years) = 4.3101

ESAL (2019) = 232,505
ESAL (2023) = 1,002,120
Difference from 2019 to 2023 = 769,615

Reference: NCEES *PE Civil Reference Handbook* > Geotechnical and Pavement > Monthly Adjustment Factor
Answer: C

SOLUTIONS

76. Refer to the *AASHTO Guide for Design of Pavement Structures*, Tables D.4, D.5, and D.6 for single-axle, double-axle, and triple-axle loads.

TRAFFIC DATA	
Initial ADT	18,000 vpd
Growth rate	2.5%/yr
Design period	20 yr
Fraction of truck traffic	0.15

AXLE LOAD DATA	
Axle load (lbf)	Number of axles
SINGLE AXLES	
4,000	100
8,000	950
12,000	1,870
22,000	60
26,000	20
DOUBLE AXLES	
6,000	510
10,000	620
32,000	100
TRIPLE AXLES	
10,000	300
18,000	400
44,000	60

Notice that the information entered in columns (A) and (C) is given for this problem.
Column B contains axle load equivalency factors found in Tables D.4, D.5, and D.6.
The entries in column (D) are the product of the entries in columns (B) and (C).
The sum of the entries in column (D) is the current total number of ESALs applied to this road in a single day.

Reference: AASHTO *Guide for Design of Pavement Structures* > Appendix D > Tables D.4, D.5, and D.6
Answer: C

SOLUTIONS

77. Using the AASHTO table for soil classification, location 1 = A-2-6, location 2 = A-7-6, and location 3 = A-7-6. Therefore, locations 2 and 3 require soil stabilization (soil stabilization is a construction technique to improve the bearing capacity of the soils).

Reference: NCEES *PE Civil Reference Handbook* > Soil Classification and Boring Log Interpretation > AASHTO Classification System
Answer: D

78. If a culvert is flowing partially full, it functions as an open channel flow meaning that critical, subcritical, and supercritical flows are all possible. When the pipe is flowing full, it functions as a pipe under pressure. Therefore, choice C is not possible.

Reference: NCEES *PE Civil Reference Handbook* > Gradually Varied Flow Channel Profiles > Froude Number
Answer: C

79. Using Manning's equation:
$$Q = \frac{K}{n} A R_h^{2/3} S^{1/2}$$

Divide by A to solve for V:
$$V = \frac{(K R_H^{2/3} S^{1/2})}{n}$$
$K = 1$ (SI units)
$S = 0.005$
$n = 0.011$
$R_h = 0.2$
$$V = \frac{(1)(0.2^{2/3})(0.005^{1/2})}{0.011} = 2.2 \text{ m/s}$$
Therefore:
Travel time = 400 m / 2.2 m/s = 182 s = 3.0 min

Reference: NCEES *PE Civil Reference Handbook* > Open-Channel Flow > Manning's Equation
Answer: C

SOLUTIONS

80. The problem requires the calculation of the net present value (NPV) for each alternative, and then obtaining the difference between the two lowest options. NPV is calculated by discounting all the future costs into the base year (0). For that, it is necessary to apply the engineering economic formulas for the annual maintenance (applied between years 1 and 9) and the rehab at year 10.

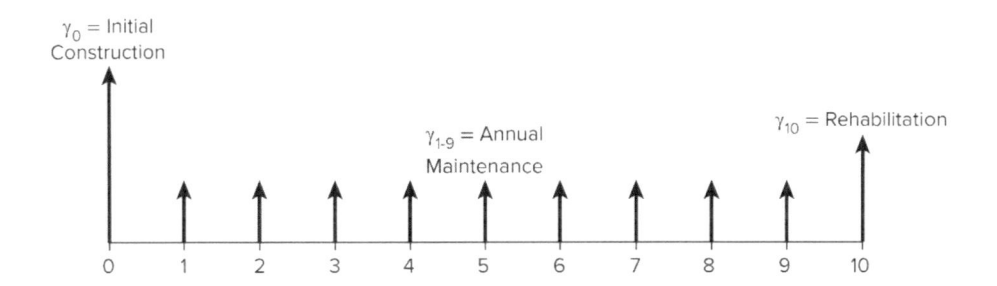

A: $1M + (P/A, $20,000, 2%, 9) + (P/F, $300,000, 2%, 10) = $1,409,334
B: $500,000 + (P/A, $50,000, 2%, 9) + (P/F, $500,000, 2%, 10) = $1,318,260
C: $1.5M + (P/A, $10,000, 2%, 9) = $1,577,861

The difference between the lowest options, A and B, is equal to $91,074 ($100,000).

Reference: NCEES *PE Civil Reference Handbook* > Engineering Economics
Answer: A

You may have taken this practice test and studied the solutions,
but what else can you do to thoroughly prepare for exam day?

Keep practicing!

One of the best study tools we offer is our Practice Portal Pro, which contains a bank of practice problems and solutions that closely mimics the NCEES' computer-based test (CBT) experience. There, you will be able to see and work through problems in the same format and style as your exam. The system will provide detailed answers so you can cross-reference your work and pinpoint where you may need improvement. Our goal with the Practice Portal Pro is to help you study smarter, not harder.

To gain immediate access to this comprehensive and time-saving study tool, download the free TotalAR app as explained at the front of this book. Use the app to scan the TAR code below and purchase the Practice Portal Pro (a $390 value).

This extremely useful tool will help you improve knowledge retention and gain a better understanding of the material.

www.schoolofpe.com